AF721464

Risks and the Anthropocene

SCIENCES

Geography and Demography, Field Director – Denise Pumain

Geography of Risk, Subject Head – Samuel Rufat

Risks and the Anthropocene

Alternative Views on the Environmental Emergency

Coordinated by
Julien Rebotier

iSTE

WILEY

First published 2021 in Great Britain and the United States by ISTE Ltd and John Wiley & Sons, Inc.

Apart from any fair dealing for the purposes of research or private study, or criticism or review, as permitted under the Copyright, Designs and Patents Act 1988, this publication may only be reproduced, stored or transmitted, in any form or by any means, with the prior permission in writing of the publishers, or in the case of reprographic reproduction in accordance with the terms and licenses issued by the CLA. Enquiries concerning reproduction outside these terms should be sent to the publishers at the undermentioned address:

ISTE Ltd
27-37 St George's Road
London SW19 4EU
UK

www.iste.co.uk

John Wiley & Sons, Inc.
111 River Street
Hoboken, NJ 07030
USA

www.wiley.com

© ISTE Ltd 2021
The rights of Julien Rebotier to be identified as the author of this work have been asserted by him in accordance with the Copyright, Designs and Patents Act 1988.

Library of Congress Control Number: 2021945744

British Library Cataloguing-in-Publication Data
A CIP record for this book is available from the British Library
ISBN 978-1-78945-041-5

ERC code:
SH2 Institutions, Values, Environment and Space
SH2_7 Environmental and climate change, societal impact and policy
SH2_11 Human, economic and social geography

MIX
Paper from responsible sources
FSC® C013604

Contents

Foreword

Risks and the Anthropocene: A Look at the Environmental Emergency

Samuel RUFAT
Institut universitaire de France, Paris, France

Since 1980, more than 1.5 billion people have died in disasters and the UN estimates that annual damages will exceed $400 billion by 2030. Global trends of increasing exposure, high inequality, rapid urban development and worsening environmental degradation raise concerns that risk levels are becoming an impediment to sustainable development. In this context, the "migration" of the notion of the Anthropocene from the earth sciences to the social sciences has led to both hopes and controversies while polarizing attention around environmental emergencies. Are we witnessing a renewal, a refoundation or a refutation of knowledge, practices and levers of action around risks and catastrophes?

With regard to "Geography of risk", books on this theme take stock of the renewal of concepts, approaches, issues and tools at the interface between risks, societies and the environment, because risks do not always translate into disasters and the increase in risks is not inevitable. It is possible to reduce the exposure of populations, to reduce their vulnerabilities, to support their adaptation or to support the resilience of societies in the face of risks that cannot be reduced or shared. Risk prevention, disaster reduction and mitigation of their effects require interdisciplinary approaches, in a subtle balance between the social demand for coordinated action at all

Risks and the Anthropocene,
coordinated by Julien REBOTIER. © ISTE Ltd 2021.

levels and the imperative of reflexive criticism, in order to deal with the substance of the problems rather than the most visible symptoms or the most fashionable solutions. This book shows that, like other such notions. “The Anthropocene, even if it is deconstructed and criticized by the social sciences, seems to function as a performative category (which tends to impose a conception of the world), more than as a heuristic category (which allows meanings to emerge around conceptions and relationships to the world)”.

While scientists’ calls to action on disasters and climate are ancient and necessary, the call to the social sciences themselves remains fundamental. This book examines the shift toward the “social sciences of the Anthropocene” by showing that the imperatives of the emergency and action are engulfed in a constant deficit of reflexivity. Julien Rebotier has brought together a multidisciplinary team of French-speaking specialists, researchers and teachers, to put their expertise into dialogue. The authors offer a critical analysis of the links between risks, the Anthropocene and environmental emergencies, without glossing over the controversies and the political and ethical issues. They show in a fine way how the emphasis on climate, earth or life sciences, when social sciences are not audible in environmental research, tends to orient both the questions and the answers in a way that is not neutral: “The environmental emergency and the thinking of the whole deny the less fortunate their very status as dominated and lock them into a double punishment: they are both the most vulnerable and those whose voice counts the least (or for whom we speak the most)”.

In a concern for rigor and pedagogy, the authors first unravel the issues and approaches of the Anthropocene to propose a critical analysis in three parts: by presenting emerging risks and emergencies, then by discussing what the Anthropocene does to the concepts, themes and approaches of risk studies, and finally by questioning the possibilities of prediction, action and reflexivity in a world of uncertainty. Each chapter can be read independently. They are organized around common issues and converge to show that “the notion of the Anthropocene intervenes mostly in parallel rather than as a complement to decades of accumulated knowledge from research and action programs on risk.... First of all, reinforcement of the naturalization implied by the Anthropocene is mentioned in many places. Our gaze is focused on the threats and diverted from the issues as well as from the genesis of environmental problems”. Their analyses are illustrated by a wide variety of examples and welcome clarifications at different scales, placing the issues in relation to the concepts dealt with in other books on this theme (adaptation, management, resilience, vulnerability, etc.), without succumbing to the effects of popularity or urgency.

With this book, Julien Rebotier and his team draw up a critical prospective assessment of studies on risks and disasters and propose to provide the social sciences with a more autonomous research agenda in the Anthropocene era. They stimulate an essential reflection for a better accompaniment of societies in the face of global change, environmental degradation and the highlighting emergencies, in the perspective of a sustainable and liveable future for all.

Introduction

Risks and the Anthropocene Moment: Social Science Issues[1]

Julien REBOTIER
TREE, CNRS – E2S UPPA, Bayonne, France

Nous dont la lampe le matin [We whose lamp in the morning],
Au clairon du coq se rallume [At the rooster's bugle is lit again],
Nous tous qu'un salaire incertain [All of us whose uncertain wages]
Ramène avant l'aube à l'enclume [Bring us back before dawn to the anvil]
Nous qui des bras, des pieds, des mains [We who with arms, feet, hands],
De tout le corps luttons sans cesse [With our whole body struggle ceaselessly],
Sans abriter nos lendemains [Without sheltering our tomorrows]
Contre le froid de la vieillesse [Against the cold of old age]

Pierre Dupont, *Le chant des ouvriers*, 1846

Collectively, we choose what we are sensitive to
and what we need to react to quickly

Bruno Latour, *Facing Gaia*, 2015, p. 396

1 This introduction has benefited from the proofreadings and comments of Florent Champy, Pascale Metzger, Catherine Meur-Ferec, Alain Musset, Patrick Pigeon and Sébastien Roux. They are warmly thanked. Thanks also to Denise Pumain and Samuel Rufat for their valuable comments on the book as a whole.

Risks and the Anthropocene,
coordinated by Julien REBOTIER. © ISTE Ltd 2021.

I.1. Risks renewed by the Anthropocene?

I.1.1. *Social science research and risk*

Social science research on risks and disasters multiplied after the Second World War. The range of approaches, concepts and methods used is extremely wide. Different currents feed a plethora of academic literature and quantities are regularly produced offering varied outlooks. Whether it is about collective or individual behaviors (Burton *et al.* 1978; Rodríguez *et al.* 2007), root causes (Blaikie *et al.* 1994; Wisner *et al.* 2012), the social construction of risks (Oliver-Smith *et al.* 2017), issues of maldevelopment or justice (Maskrey 1993), or readings of societies and their relationships to environments (Douglas and Wildavsky 1983; Theys and Fabiani 1987; Beck 1992), the intelligibility of risks and their management from the social world defines a vast and diverse knowledge domain.

Note the considerable deployment of initiatives of all kinds (research, studies, policies) on the risks of disasters, on the mechanisms that lead to them, on the ways to identify them and even to manage them (Pigeon and Rebotier 2016). So many efforts question the scope of research in the world that it takes as its object and of which it is part (Ribot 2019). The history and sociology of science show this (Bourdieu 1975; Bonneuil and Pestre 2015), and research on risk bears witness to it: there is no causal link between knowledge and action. Research is one way to produce knowledge (Mercer 2012). It coexists with other interests that are foreign to it and that weigh on it as well as on the arbitrations of action. Research is shaped by the world in which it takes place.

I.1.2. *The Anthropocene: a new moment for thinking about risks?*

Today, the notion of the Anthropocene is the focus of attention for environmental concerns. However, its content and temporal boundaries are still under discussion and controversy. Consequently, similarly to the French version of the book, the Anthropocene will not be capitalized throughout the book[2]. This editorial choice has not been imposed on the authors, but each of them has been careful to specify the preferred meaning when using the term. Whether it is a question of the more restricted geological meaning, of the craze that the notion arouses, of its effects in the political and scientific spheres, or of the reflection of a broader context for thinking, living and

2 In the English language, however, it is usual to use a capital "A," which does not prevent some authors from referring to the "Anthropocene" as an ontological moment (Clark and Yusoff 2017, p. 8), a broader conjuncture of ways of thinking, living, and doing the world, in the same way that we understand the Anthropocene moment in this book.

making the world, the effort to clarify the meanings of a term that remains polysemous is recalled by the use of Anthropocene[3], or even Anthropocenes.

Engaged in this work of elucidation, the social sciences provide information on risk situations and their meanings (material, symbolic, political, economic) in their contexts. In this way, they shed light on the social world and the contexts in question, and on the moments to which the risk situations and their analysis correspond. Reducing the consequences of disasters is a legitimate objective for research. But the social sciences also have the mandate to produce intelligibility, in this case through the undertaking of risk situations. In this sense, risks can be reconsidered in the light of an Anthropocene "moment", in the same way as other phenomena or issues such as globalization, justice or resilience (Chandler *et al.* 2020). The Anthropocene would introduce a "remarkable sequence" for renewing the understanding of life on Earth due to new articulations between living beings and the biogeochemical mechanisms of the planet. It remains a moment to be defined.

It is difficult to account for the diversity (and even contradictions) and the finesse of knowledge on risks. It is also difficult to introduce the notion of Anthropocene without arbitrating between its meanings, and between the terms used. Sequence or geological epoch? Living beings rather than human beings, whereas the human race often locks the discussion into the opposition: humanity *versus* the rest of the world. Choosing to use the category of humanity is a shortcut of thought that is rarely used in a measured way. It carries laudable ideals as well as obliterating abysmal inequalities. Finally, is the Earth correct to refer to a system and conditions that allow the living being to be accommodated? The planet Earth, then, unifies yet certainly oversimplifies the complexity of biogeochemical processes that we are now sure have a global history since their unprecedented trajectory is incomprehensible without human influences.

The Anthropocene is a 21st-century notion, marked by the seal of the physical and earth sciences. Like many other concepts, it has been taken up by the social sciences and is the subject of controversial, sometimes contradictory and virulent debates. Once the "Anthropocene event" (Fressoz and Bonneuil 2015) has passed through the stupefaction of its possible implications and the vagueness of its definition, which goes hand in hand with a certain anxiety (can we really measure it?), it is appropriate to return to the way this event forms a turning point to the ordinary and significant course of social relations and events; in the modes of production, of living, of circulating; in the complexity of the interactions that are established between humans and non-humans. To give an account of an intellectual, institutional, scientific and political conjuncture at work in the opinion as much as in more academic or programmatic registers sends us back to the complexity of a

3 Not capitalized in the French version of the book.

moment. In this sense, the Anthropocene moment corresponds to a historical period that is characterized by ways of thinking, arguing and acting, both collectively and individually, particularly with regard to the attention given to the environmental issue in our societies. The Anthropocene as a moment refers as much to the possible objectifications of an unprecedented sequence of relations between societies and environments as to the need to reconsider certain reference points and major structuring narratives of a modernity that is emerging from the 20th century bled dry: the great divide (between nature and culture), the status of science, the trajectory of progress, the mastery of uncertainty, security, etc.

The Anthropocene will be discussed in this introduction as a singular moment that seems to be taking shape, but it is indeed risk research, in its diversity and from the social sciences, that is at the heart of this book. In many respects, the Anthropocene moment remobilizes old questions about risk: relations between societies and environments, solidarity and inequalities, the hegemony of the naturalistic approach, the political nature of environmental issues, etc. Interrogating the knowledge produced by the social sciences on the notion of the Anthropocene thus makes it possible to point out fundamental debates, often hidden behind simple reformulations.

I.2. Defining the Anthropocene: ways of seeing, ways of thinking

I.2.1. *A global framing by the Earth sciences*

In the early 2000s, geochemist Paul Crutzen helped popularize the notion of the Anthropocene (Crutzen and Stoermer 2000). The rapid introduction of the notion into the global science arena was partly orchestrated from the International Geosphere-Biosphere Programme – an international initiative led by the International Council for Science (ICSU) – where Crutzen served as vice-president. The IGBP aims to coordinate research at global and regional scales on the earth system, on the biogeochemical interactions that constitute it, and on the interactions maintained with societies. The IGBP is part of the trend toward the globalization of environmental issues, toward the consolidation of a planetary, atmospheric and climatic framework, then more broadly biogeochemical, and toward the train of major scientific initiatives driven by international governance – through the United Nations system. Thus, the creation of the IGBP in 1987 followed that of the World Climate Research Programme in 1979. Concerns about climate extend to global change in general in order to measure human impacts on the environment, from what would become the Earth sciences, which are well suited to address changes in the earth system. In 1988, the Intergovernmental Panel on Climate Change was formed, laying the foundations for a type of governance of climate and environmental issues already analyzed by the social sciences (Dahan and Aykut 2015).

The Anthropocene postulates that humanity's influence on the geosphere and biosphere is such that it can define a specific geological epoch that would bear its name (Crutzen 2002). The recognition of human influence on the Earth is not new (Marsh 1865; Crutzen 2007) and, through a form of human ecology, it even constitutes one of the foundations of geographical questioning (Robic 2006, pp. 28–29). But the recent success of the notion can be understood in two ways. It is due to the intensity and unprecedented nature of interactions between societies and environments (Gemenne *et al.* 2019), as well as to the conditions in which scientific knowledge is produced, to the structuring of a globalized intellectual climate that weighs heavily on the practice of research (Castree *et al.* 2014), and to the place of the environmental question within globalized societies (Smith 2010). These two bundles of elements, namely the physical reality of unprecedented consequences of human–environment interactions, on the one hand, and the social configurations that accompany them, on the other hand, constitute the Anthropocene moment. The latter is of interest to the reflections gathered in this work on risk assessment and risk management.

I.2.2. *From controversies on the boundary...*

I.2.2.1. *The steam engine, the Industrial Revolution and the Great Acceleration*

Proposing the Anthropocene as a new geological epoch requires compliance with the rules of the International Commission on Stratigraphy, the body within the International Union of Geological Sciences that oversees the standardization of a geological time scale. Designating the boundaries of a geological epoch requires two conditions. It is necessary to be able to locate the traces of a global event left in stratigraphic materials (rocks, ice, sediments) and to associate them with other stratigraphic markers indicating changes in the whole earth system. Duly localized traces constitute a "golden spike"[4].

For Crutzen and Stroemer (2000), it is the increase in the amount of CO_2 found in the atmosphere (visible in ice cores) that marks the beginning of the Anthropocene at the end of the 18th century. This period corresponds to the beginnings of the industrial revolution. The authors link it to the introduction of the

4 In the absence of a "golden spike", especially in the most remote geological ages, the boundary can be established by scientific consensus on the basis of less definitive stratigraphic evidence.

steam engine patented by James Watts in 1784. Thanks to the multiplication of the data collected, a period of "great acceleration" after the Second World War has been identified more precisely (Steffen *et al.* 2007). The malady was diagnosed, tracing the horizon of a techno-scientific and moral research agenda supposed to respond to the major dysfunctions that this evolution has partly caused:

> A daunting task lies ahead for scientists and engineers to guide society towards environmentally sustainable management during the era of the Anthropocene. This will require appropriate human behaviour at all scales, and may well involve internationally accepted, large-scale geo-engineering projects, for instance to "optimize" climate. (Crutzen 2002, p. 23)

Despite the confidence displayed, evidence is lacking to close the stratigraphic debate on the Anthropocene. Above all, underlying ways of seeing the world clash across such categorization, underscoring the interest of axiological controversies in addition to metrological ones:

> Care is needed to ensure that the dominant culture of today's scientists does not subconsciously influence the assessment of stratigraphic evidence. (Lewis and Maslin 2015, p. 173)

I.2.2.2. *A range of possible hypotheses*

Geological time is marked by changes in the state of the Earth. Between the end of the Pleistocene (11,700 years ago) and the 1960s, several events are strong candidates as anthropogenic signatures of geologic epoch change. Lewis and Maslin (2015) explore several hypotheses and select two.

The great fires associated with the Pleistocene mass extinctions are discarded, their traces not being explicitly global enough. The stratigraphic evidence of the beginnings of agriculture and the upheavals of the Neolithic are not considered sufficiently synchronous. The process of industrialization and then the industrial revolution bequeath diverse and temporally dispersed markers. On the other hand, 1610 and 1964 are favored by authors as potential starting points of the Anthropocene.

For 1610, the consequences of the discovery of the Americas by the Europeans at the very end of the 15th century led to the global dissemination of pollens of many species and to the massive sequestration of carbon. In a few decades, the genocidal impact of the bacteriological shock on the Amerindian populations led to the death of 90% of the indigenous population and the rewilding of 50 million hectares of forest, wooded savannahs and grasslands, for lack of manpower to work them. Considerable amounts of CO_2 were captured, removed from the atmosphere, as indicated by the ice cores that report the date of 1610 (Lewis and Maslin 2015, p. 175). For 1964, in the context of the "Great Acceleration," it is a spike in the concentration of ^{14}C (a radioactive isotope of carbon) in ice and trees that wins favor with Steffen *et al.* (2015). The ^{14}C spike evidenced in the stratigraphic material was linked to a frequency of nuclear explosions on the Earth's surface unmatched by any other time period.

Each hypothesis reflects an image of societies and their relationship to the environment. The date of 1610 emphasizes the domination and exploitation of resources. The 1964 date points to the concord of an international governance capable of banishing – or at least reducing significantly – the use of nuclear weapons. The fact remains that there is no consensus on the stratigraphic evidence and, above all, that it is only a matter of conforming to the requirements of geology. As such, the axiological controversies surrounding the Anthropocene carry more meaning – and are of greater interest – to the social sciences than the debates on the stratigraphic classification that presides over its identification from the Earth sciences.

I.2.3. *... to the strengthening of meta-narratives*

As a result of the disciplinary divisions, the truth of geology is not the same as that of the social sciences. Framing effects are at work here. Under cover of the authority of the discourses produced by legitimate sciences, we can convey ways of seeing the world. It is the role of the social sciences to identify the meanings produced, to identify the sources of the arguments on which they are based, as well as the interests, the expectations or the benefits linked to social positions. Numerous works contribute to exploring the social scope of the controversies surrounding the Anthropocene (Lorimer 2017), to making explicit the images of the world conveyed (Bonneuil 2015), and even to considering the agenda of autonomous research on the subject (Lövbrand *et al.* 2015; Davis and Todd 2017).

In this respect, the identification of major narratives conveyed by the Anthropocene is instructive of the possible contributions of social sciences (Bonneuil 2015):

– Earth sciences introduce a naturalistic narrative, which separates man and nature, by giving science and technology the means to sound the alarm and to find solutions. The advent of a global and homogeneous environmental conscience justifies that problems and solutions draw from the same source (Crutzen and Stroemer 2000);

– the postnature narrative does not recognize the modern separation between nature and culture. It comes into play to the extent that it recognizes that the outcome of actions is beyond the control of the will alone. We humans "become geology" (Latour 2015) and the emerging reflexive subject must acknowledge this (Beck 1992). The great divide between nature and culture blurs to define a common hybrid, of humans and non-humans, able to identify the environmental question as superior and ultimate;

– the ecocatastrophic narrative has been a recent bestseller (Servigne and Stevens 2015), while at the same time reviving the apocalyptic reflections of the post-Second World War era (Anders 1995) or those dealing with the autonomy of technology (Mumford 1950). From the limits of the planet to the prospect of collapse, it is impossible for societies to think of themselves as being alone in the maneuver. The capacity for action is shifted toward a kind of ecomodernism that would allow us to act in a more relational way on/with the environment (of which societies are part);

– the ecomarxist narrative applies the grid of reading historical materialist antagonisms (Keucheyan 2014; Moore 2017a), opposing classes and interests. The Anthropocene moment is the contemporary theater of the contradictions of capitalism, especially those that confront it with the depletion of resources in a finite world. The environmental crisis is the manifestation of an asymmetrical globalization of the economic world, reactivating the themes of inequality, exploitation and domination;

– the feminist political ecology narrative links the mechanisms of economic exploitation that transpire during the Anthropocene to other strands of domination. Issues of gender, community, racialization, sexual orientation, or faith, from an intersectional perspective (Gandy 2015; Yusoff 2018), weave the differentiated fabric of environmental crisis.

The work of contextualizing and explicating the grand narratives that accompany possible Anthropocenes also concerns the discourses held by the social sciences themselves (Lövbrand *et al.* 2015). Social science knowledge is always produced in the plural, under social conditions that need to be identified, both within the scientific field and in the social world (Bourdieu 2001).

I.2.4. *Thinking about Anthropocenes from the social sciences*

During the last decade, the growing production of work by the social sciences on the Anthropocene has been "foundational". Collective articles seek to "reconceptualize" the Anthropocene – so as not to leave it to the Earth sciences? To defend competing interpretations of the social world? (Palsson *et al.* 2013; Lövbrand *et al.* 2015; Brondizio *et al.* 2016). Some elements of reflection are recurrent. This is the case of Anthropocene thought as an opportunity to break with the modern division between nature and culture. But very distinct positions clearly identify very different currents, sometimes in obvious opposition:

– the stream most compatible with the framing of the Anthropocene by the Earth sciences claims a joint research agenda (Brondizio *et al.* 2016). We find co-publishing authors who share a form of universalization of processes (Steffen *et al.* 2007) and agree on the identification of thresholds (*tipping points*), the place of numbers in a generalized equivalence and the evaluation of limit values with a global scope (*planet boundaries),* or an approach based on scenarios. However, the potential coupling of "social sciences and physical sciences" (Castree *et al.* 2014) is only a fraction of what the social sciences can contribute. Integration is not questioned when it suggests that the sum of knowledge necessarily leads to better knowledge... which is not self-evident, however;

– another current claims a more interpretative perspective that explores the possible meanings of the Anthropocene for the social world through a new form of weaving on (and with) the Earth (Tsing 2017). Humans are becoming aware of a common, interdependent humanity through the multiple, complex, random and not entirely predictable interactions with living things and the rest of the geochemical dynamics. The Gaia hypothesis is taken up by Lovelock and Margulis (1974) through the Anthropocene and this new era that forces Earthlings to cohabit with others, human and non-human, "under the authority of a power without political institution yet assured" (Latour 2017, p. 115). The reflexive human subject is no longer the pivot of an anthropocentric existence. It is the "assemblages of organic species and a biotic actors that make history" (Haraway 2016, p. 76) and create new familiarities, extended kinships, toward another form of shared world (as exemplified by the COVID-19 pandemic in 2020). Proponents of this current assess the complex "web of life", which is "unruly, rebellious, and has a way of continually upsetting the best laid plans of states, of capitalists, of scientists and engineers" (Moore 2017c, p. 177);

– a third current breaks more clearly with the Earth sciences (Lövbrand *et al.* 2015) as with the proponents of a "posthuman" thought (which moves from the human condition to that of Earthlings, human and non-human). Vectoring a form of pagan spirituality, the fable of the "We-common-earthling" (or humanity) is not convincing (Hornborg 2017). For this latter, yet very composite, current, blindly universalizing thought disarms the critique of capitalism, the springs of injustice and forms of domination (Malm and Hornborg 2014). In order to shed light on the matrix genesis of the Anthropocene, authors have introduced new terms such as "capitalocene" (Moore 2017a, 2018), pointing to globalized capitalist exploitation, or "plantacionocene" (Ferdinand 2019), denouncing the colonial fact. In both cases, the idea is not to reduce the Anthropocene to a material consequence of capitalism or colonialism but to extend it to a relationship to the world, to ways of relating to all elements of the "web of life," to an imaginary penetrated by multifaceted and invasive fronts of appropriation that place all things in equivalence (Moore 2003). The perspective of a "world-ecology" reveals the divergence of elements and their complex, reciprocal relationships that connect – and make – the world without obliterating the inequalities, asymmetries, contradictions and antagonisms of a more materialist analysis (Peet and Watts 1996; Castree *et al.* 2014; Davis and Todd 2017).

Several lessons can be drawn from social sciences readings of the Anthropocene. These echo the persistent and structuring variety found in risk studies. Bringing together debates on the Anthropocene and reflections on risk allows us to confirm, complete and even deepen certain recurring obstacles and challenges in the understanding and management of disaster risks, as well as of the environmental question more broadly.

I.3. Risks and social sciences: well-identified obstacles and challenges that continue to be debated

I.3.1. *The blind spot of development*

Ben Wisner defines "disaster studies" as:

> A broad interdisciplinary attempt to understand the causes and consequences of events that cause sufficient harm and loss that assistance is required from people and/or institutions unaffected, whatever size of the group and area affected. (Wisner 2019, p. 48)

This very open definition of a composite field leaves unresolved the questions that concern "the causes and effects of events that lead to damage". Since the second half of the 20th century, a growing part of research has been devoted to the denaturalization of so-called natural hazards (O'Keefe *et al.* 1976). But the social processes put forward are based on different, sometimes antagonistic, visions of the social world and relationships with the environment. It is sometimes a question of being attentive to the behavior of individuals, sometimes to the response capacities of communities, sometimes to the effects of framing social structures, among many other approaches. Alongside this polyphony of social sciences around disaster risks, the understanding and treatment of hazards continue to be a major focus of attention, effort (and resources) of research and social demand:

> Four decades of academic literature on disasters (e.g. Baird *et al.* 1975; Maskrey, 1989; Oliver-Smith, 1994), backed up by a profusion of practitioners' reports from the field (e.g. Anderson and Woodrow 1989; Heijmans and Victoria 2001), have shown that disasters deeply reflect failed or skewed development. Considering vulnerability to natural hazards through the sole lens of potential damage created by rare and extreme natural phenomena is a remnant of a paradigm that has been completely up-ended. (Wisner *et al.* 2012, p. 11)

In a literature that purports to be intermediate between academia and action, despite an observation that is now half a century old, the challenge remains:

> If disaster risk is an endogenous indicator of a flawed development model, then progress toward the policy goal of disaster risk reduction will depend on a transformation of that model. If the world is to survive beyond the middle of the 21st century... it will be necessary to make conclusive progress on the path that has been least followed under the Hyogo Framework for Action and to develop a new approach to disaster risk management. (GAR 2015, p. 39)

However, the GAR (Global Assessment Report on Disaster Risk Reduction) does not engage in the programmatic field. Yet the Sendai Framework for Action (2015–2030), which postdates the Hyogo Framework, does not meet the need to turn the correction of misguided development into the key driver of risk management either (Wisner 2016, p. 32). In theory, the relevance of social science input is no longer in dispute. In practice, misunderstandings and confusions persist about their place, scope and impact in research and management (Ribot 2019). This discrepancy should be seen in the light of the apparent contrast between the many efforts made to understand risks, on the one hand, and the observation of a simultaneous increase in damage, on the other hand (White *et al.* 2001).

I.3.2. *Risks and the Anthropocene: updating or evacuating social science questions?*

I.3.2.1. *Old themes that are still relevant*

Given both the importance of the Earth sciences in framing the Anthropocene and the discrete consequences of critical social science work in the field, one might ask whether the Anthropocene is advancing a meaningful research agenda on risk, or whether, on the contrary, it is helping to make certain social science productions incidental. The themes found around the Anthropocene are illustrated by these five examples:

– Knowledge: there is not a single piece of knowledge, but varying forms of knowledge, which may or may not be scientific, may not always be compatible and may sometimes be in competition to aid understanding of the "causes and consequences" of harmful events (Mercer 2012). The social sciences are equipped to decipher the meanings encapsulated in the different acceptations of the notions mobilized (Demeritt 2001; O'Brien *et al.* 2007; Hulme 2009; Bonneuil 2015). Knowledge never evolves alone, independently of the social world in which it is produced and conveyed.

– Technocracy and dogmatism: the decision founded in science remains a political arbitration. On the one hand, the practice of science and the legitimacy it enjoys – not to mention the heterogeneity of the scientific field – are a matter of social relations and therefore of politics (Bourdieu 1975). On the other hand, recourse to science to decide is always selective. It depends on the moment, on the context or on the underlying conceptualizations that presume a reading of the world. If they are not made explicit, they are ideology. Pretending to avoid arbitrations on the grounds that geology or atmospheric sciences would know neither morality nor subjectivity is an act of faith.

– Scientific polyphony: for risks as for the Anthropocene, the diversity of social science productions is considerable. This introduction has mentioned work on both risks and the Anthropocene. In both cases, there is no consensus on the analysis of the social dimensions of environmental issues, and there is a wide range between utilitarian reflections, in line with Earth science approaches, on the one hand, and the analysis of the environmental dimensions of the social, on the other hand. This diversity, characteristic of the epistemology of the social sciences (Passeron 1995), enriches interpretation but can feed confusion and so be detrimental to such environmental research.

– Interdisciplinarity and integration: it is now commonplace to denounce the fragmentation of research into silos, particularly in the case of subjects such as risk. We distinguish two forms of integration called for in response. One concerns the conditions of research; the other concerns the decompartmentalization of disciplines around the society–environment relationships that are the hallmark of geography and constitute one of its historical markers.

– The power of capture of one legitimate research sector over another is known (Castree *et al.* 2014). Some promote the broad field of "sustainability sciences" (Lorimer 2017) or embrace the unifying dynamics of "environmental sciences" with the Anthropocene (Brondizio *et al.* 2016). Others denounce it, see the opportunity to think differently (Palsson *et al.* 2013) and resist falling into line with a poorly controlled framing of research. In particular, the debate stumbles over modeling or indicators, in fact over reluctance to put an irreducibly diverse social world into functional equivalence (numerical or not) (Wisner 2016). Extracting the figure from its context also cuts the data from its meaning (Rebotier *et al.* 2019) and contributes to dehistoricizing the analysis (Moore 2003).

– Historically, risk studies have endorsed the separation between nature and culture (between hazard and vulnerability). This is particularly visible in the use of conceptual models that struggle to articulate physical processes and social mechanisms in a complex and explicit way (Pigeon and Rebotier 2016; García Acosta and Musset 2018). The Anthropocene moment may offer an opportunity to think about the post-great divide between nature and culture. But integrated ways of considering the reciprocal and non-deterministic relationships between societies and environments were already among the founding debates of geography at the turn of the 19th and 20th centuries (Berdoulay 2008). The trajectory of the notion of adaptation is an illustration of this (Rebotier 2020), and calls for more integrated research on the environment, at least in Francophone geography, and predates the Anthropocene moment (Mathieu and Jollivet 1989).

– Transformations of the social: the theme of social change as a practice has sensibly receded in the research agenda (Castel and Martin 2012). Regardless of what it is called, (transformative development (Lavell and Maskrey 2014); adaptive transformation (Pelling 2010)), social change is either reduced to the inoffensive perimeter of small reformist steps (Pahl-Wostl 2009) whose modest consequences are continually promised to become significant; or it falls outside the scientific field, on the grounds that it would cover an ideological bias unacceptable in the majority representation of science (Metzger and Robert 2015). Research copes differently with this normative tension. Critical analyses identify asymmetrical development or structuring inequalities as obstacles to effective, sustainable and just improvements in risk conditions (Lavell and Maskrey 2014). The same criticisms are, nevertheless, blunted in official documents to preserve the consensus required by international institutionalization (GAR 2015). Critical thinking seems neutralized between its inclination to say things and a form of inability to change the world. In spite of the impasse, some people take up the issue of social responsibility in two ways. On the one hand, they assume that changing the world (not just making it work or describing it) is part of the mandate of research (Robbins 2004); on the other hand, they engage with other knowledge producers and with social sectors (other than those composed of scientists) that are more action-oriented (Mercer 2012; Shaw 2013). On the subject of the Anthropocene, should we dismiss the critique of capitalism on the grounds that its postponement is unrealistic (Latour 2017)? Along the narrow path that combines critique of capitalism and social transformation, some explore a movementist, collective option (Wisner 2019), which relies on critical initiatives and claims interdependencies as pledges of liberation (Moore 2017b).

Already known themes are reformulated and take on the appearance of a new terminology without any impression of significant progress. The apparent stagnation legitimately raises questions about a form of neutralization of social science questions and reflection. Clearly, the weak consequences of social science knowledge on risks and the environment would make it non-discriminating, without effect; in short, non-significant and almost useless in the wider debates (which go beyond corporations and disciplinary arenas), despite a great deal of valuable work, innovative proposals and ambitious approaches. A brief review of the difficulties in considering the issues of power around environmental problems, particularly in French geography, is enlightening in this respect.

I.3.2.2. *A still difficult introduction of the environment in society*

From the ranks of French geography to the arenas of globalized research, the status of the environmental question has been strongly conditioned. The Anthropocene moment seems to consolidate and even deepen the framings more than it offers real opportunities to reduce, circumvent, transform or overcome them. And this is certainly less a problem of the quality of the work than of the social conditions in which social science knowledge on risks and the environment is elaborated, produced, relayed, legitimized and appropriated (Metzger 2017).

In France, the 1990s and 2000s marked a resistance to thinking about the environment as a social issue within a geography understood as a social science (Kull and Batterbury 2017). Environmental concerns were pushed to the margins of the discipline (Pelletier 1993; Rossi 2001; Brunel and Pitte 2010), whether they blurred the more classical software of social struggles of industrial societies or compromised the scientificity of the study of natural environments that were more or less open to interactions with societies. This reluctance of some geographers to give a full place to social and political issues in environmental problems is at once symptomatic, blinding and declining:

– It is symptomatic of a modern world cornered at the threshold of upheavals it helped cause, and unable to think of itself simultaneously as an element taken in these upheavals. Tetanized by their power and its consequences which put them under obligation, modern actors evade their responsibility (Latour 1991) and try to disqualify their denigrators, who support the entry into political debate of an eminently social environmental question. The rise of the ecological question in politics, which has been growing in France since Jacques Ellul (1912–1994),

Bernard Charbonneau (1910–1996) and Serge Moscovici (1925–2014), is referred to as the excesses of obscurantism – the Amish model and the oil lamp – or to the fascist threat – ayatollahs and other green Khmers (Ferry 1992; Brunel and Pitte 2010).

– It is blinding, because it masks the efforts, already mentioned, of fine integration of the natural and the social. Thinkers of the milieu[5], such as Reclus or Kropotkin (regularly mobilized today in literature on the Anthropocene), put forward an idea that is not exteriority to the social world, but rather the embodiment of reciprocal relations between societies and a natural environment (in the sense of what surrounds), whether living or not. The notion did not resist neither the regional approach, nor later the spatial turn of quantitative geography, while abandoning nuances and complexity of the idea of milieu (Robic 1992). During the 1970s, the notion of environment became institutionalized as the systemic study of a complex of interdependent societies and natural environments in the form of a geosystem structured in geography (Bertrand 1968). At the same time, national interdisciplinary scientific programs sought to combine social and physical approaches to an environmental issue that was emerging as a social issue (Tissier 1992). In spite of the resistance to make room for social questions, these two precedents (and there are many others, in a wide range from Jean Tricart to Augustin Berque) show how much what will be called environmental geography only gradually, laboriously, painfully integrates the social question.

– Finally, this resistance promises to be on the decline insofar as the stigma of a major split in the discipline (between physical geography and human geography) is pushing younger generations to defend a programmatic horizon for a radically integrated "environmental geography" (Chartier and Rodary 2016). Another informal group formed in 2016, called Cynorhodon, is, for example, behind the *Critical Dictionary of the Anthropocene* (Cynorhodon 2020). The spirit of the times is one of permeability between disciplines, but also between questionings, from the plural register of environmental humanities (Choné *et al.* 2016) to the anchoring in critical geography of a physical practice of the discipline (Di Mauro 2015; Dufour 2015; Dufour and Lespez 2019). In the wake of this new breath of environmentalism on renewed practices and issues of geography, we find the emergence in France of the scientific current of political ecology (Gautier and Benjaminsen 2012). The specificities of the corporation, the institutional structuring of the community, the proximity between geography and public policy or the development of related problematic fields (such as environmental justice) certainly contribute to explaining the late introduction in France of a current formulated in the 1970s and well established in English-language, but also Spanish-language, environmental geographies (Kull and Batterbury 2017).

5 As an assemblage of physical and human features involving different spatial scales.

Thus, the evolution of French geography with respect to the environmental question is not induced by an Anthropocene moment. Rather, it is part of the long history of a disciplinary trajectory, its structuring narratives and competing paradigms and approaches. The Anthropocene moment and its environmental emergency today (like the awareness of these issues in the 1970s) sharpen the tensions within a discipline that is constantly working, recomposing and repositioning itself. The "missed appointment" (Chartier and Rodary 2016) of an integrating environment for geography at the beginning of the 20th century could today present a new face.

This is the meaning of the status of the social in the environment. Natural, ecological, physical conditions or planetary limits are not, for the social sciences, the cardinal terms of the environmental problem. Rather, through the classic but robustly updated idea of the co-production of nature (Braun and Castree 1998; Ekers and Loftus 2012), it is social and political conditions that are central to ways of regulating a shared, safer and more sustainable world. The environmental problem does not surround the social world. It is a product of it, the result of social relations and interactions between societies and their environments (Ribot 2014).

Seen from the South, what difference does it make to interrogate risks in the Anthropocene? What new things does the Anthropocene introduce for those whose world has already collapsed or whose inequalities are already difficult to bear (Bouisset *et al.* 2018)? It has been known (read and written about) for decades that development (a form of social relations and relationships with the environment) constitutes the matrix of the greatest social challenges – with multiple environmental variations. What remains is to put relational intelligence at the service of one of the fundamental mandates of risk studies: to practically and sustainably reduce the risks of disaster by pointing out and acting on the multiple causes and on the differentiated consequences of harmful events.

I.4. Insights on risks in the Anthropocene era

Based on the Anthropocene moment, nine chapters have been proposed for the study and management of risks by providing elements of response in three parts, posed as three questions. The first part, "Toward new risks?", discusses types of risk (coastal risks, fires, urban climate) that have emerged or been renewed by the Anthropocene context. The second part, "Recompositions for the study and management of risks?", evaluates what the Anthropocene moment is doing to the classic themes of risk studies: the risk society, the management of disasters or prevention policies that are always the mark of attributes of power. The third part, "What consequences for a changing modernity?", takes a step back from the

Anthropocene moment and draws the (at least potential) contours of future worlds. It includes an ethnography of flood prediction systems in Europe, an analysis of the shift to a world of uncertainty and a practical philosophical examination of an old challenge for action: the necessary action in a radically uncertain world.

Each contribution is positioned in its own way around common problematic orientations, recalled by this general introduction. It is possible to navigate between the autonomous chapters because of explicit cross-references throughout the text and to the general index of the book.

These reflections all feed questioning in the social sciences. They bring together authors who, for the most part, are involved in risk studies. At the heart of the book, highlighted by reflections on the Anthropocene, we find risks as research objects, as supports and markers of the social and as modalities and manifestations of interactions between societies and environments. The general perspective assumes a form of déjà vu by being part of several decades of social science research on risks. The notion of the Anthropocene, despite its success, should neither be used to neglect the critical knowledge already produced on risks, their study and their management, nor to elude the obstacles already identified. The introduction of the notion of the Anthropocene into the debate on risk thus serves as a revelation of certain salient points, or as a moment of reference for drawing up forward-looking assessments.

I.5. References

Anders, G. (1995). *De la bombe et de notre aveuglement face à l'apocalypse*. Titanic, Paris.

Beck, U. (1992). *Risk Society: Towards a New Modernity*. Sage Publishing, London and New York.

Berdoulay, V. (2008). *La formation de l'école française de géographie (1870–1914)*. Éditions du CTHS, Paris.

Bertrand, G. (1968). Paysages et géographie physique globale. *RGPSO*, 39(3), 249–272.

Blaikie, P., Cannon, T., Davis, I., Wisner, B. (1994). *At Risk: Natural Hazards, People's Vulnerability and Disasters*. Routledge, London.

Bonneuil, C. (2015). The geological turn. Narratives of the Anthropocene. In *The Anthropocene and the Global Environmental Crisis: Rethinking Modernity in a New Epoch*, Hamilton, C., Gemenne, F., Bonneuil, C. (eds). Routledge, London.

Bonneuil, C. and Pestre, D. (2015). *Histoire des sciences et des savoirs. Le siècle des technosciences*, Volume 3. Le Seuil, Paris.

Bouisset, C., Clarimont, S., Rebotier, J. (2018). Résilience et prévention des désastres (dossier). *Vertigo*, Special edition 30 [Online]. Available at: https://doi.org/10.4000/vertigo.19331 [Accessed 6 October 2020].

Bourdieu, P. (1975). La spécificité du champ scientifique et les conditions sociales du progrès de la raison. *Sociologie et Sociétés*, 7(1), 91–118.

Bourdieu, P. (2001). *Science de la science et réflexivité*. Raisons d'agir, Paris.

Braun, B. and Castree, N. (eds) (1998). *Remaking Reality. Nature at the Millenium*. Routledge, London.

Brondizio, E.S., O'Brien, K., Bai, X., Biermann, F., Steffen, W., Berkhout, F., Cudennec, C., Lemos, M.C., Wolfe, A., Palma-Oliveira, J., Chen, C.-T.A. (2016). Re-conceptualizing the Anthropocene: A call for collaboration. *Global Environmental Change*, 39, 318–327.

Brunel, S. and Pitte, J.-R. (2010). *Le ciel ne va pas nous tomber sur la tête*. JC Lattès, Paris.

Burton, I., Kates, R.W., White, G.F. (1978). *The Environment as Hazard*. Oxford University Press, Oxford.

Castel, R. and Martin, C. (2012). *Changements et pensées du changement*. La Découverte, Paris.

Castree, N., Adams, W.M., Barry, J., Brockington, D., Büscher, B., Corbera, E., Demeritt, D., Duffy, R., Felt, U., Neves, K., Newell, P., Pellizzoni, L., Rigby, K., Robbins, P., Robin, L., Bird Rose, D., Ross, A., Schlosberg, D., Sörlin, S., West, P., Whitehead, M., Wynne, B. (2014). Changing the intellectual climate. *Nature Climate Change*, 4, 763–768.

Chandler, D., Grove, K., Wakefield, S. (2020). *Resilience in the Anthropocene: Governance and Politics at the End of the World*. Routledge, London.

Chartier, D. and Rodary, E. (eds) (2016). *Manifeste pour une géographie environnementale*. Presses de Sciences Po, Paris.

Choné, A., Hajek, I., Hamman, P. (2016). *Guide des humanités environnementales*. Septentrion, Paris.

Clark, N. and Yusoff, K. (2017). Geoscoscial formations and the Anthropocene. *Theory, Culture & Society*, 34(2–3), 3–23.

Crutzen, P.J. (2002). Geology of mankind. *Nature*, 415(23) [Online]. Available at: https://doi.org/10.1038/415023a [Accessed 8 May 2020].

Crutzen, P.J. (2007). La géologie de l'humanité : l'Anthropocène. *Écologie & politique*, 34(1), 141–148.

Crutzen, P.J. and Stoermer, E.F. (2000). The "Anthropocene". *IGBP Newsletter*, 41, 17–18.

Cynorhodon (2020). *Dictionnaire critique de l'anthropocène*. CNRS Éditions, Paris.

Dahan, A. and Aykut, S. (2015). *Gouverner le climat ? Vingt ans de négociations internationales*. Presses de Sciences Po, Paris.

Davis, H. and Todd, Z. (2017). On the importance of a date, or decolonizing the Anthropocene. *ACME: An International Journal of Critical Geographies*, 16(4), 761–780.

Demeritt, D. (2001). The construction of global warming and the politics of science. *Annals of the Association of American Geographers*, 91(2), 307–337.

Di Mauro, S. (2015). Postface. In *Espace et rapports de domination*, Clerval, A., Fleury, A., Weber, S., Rebotier, J. (eds). PUR, Rennes.

Douglas, M. and Wildavsky, A. (1983). *Risk and Culture: An Essay of the Selection of Technological and Environmental Dangers*. University of California Press, Berkeley, CA.

Dufour, S. (2015). Sur la proposition d'une géographie physique critique. *L'Information Géograhique*, 79, 8–16.

Dufour, S. and Lespez, L. (2019). Les approches naturalistes en géographie, vers un renouveau réflexif autour de la notion de nature ? *BAGF*, 96(2), 319–343.

Ekers, M. and Loftus, A. (2012). Revitalizing the production of nature thesis: A Gramscian turn? *Progress in Human Geography*, 37(2), 234–252.

Ferdinand, M. (2019). *Une écologie décoloniale – Penser l'écologie depuis le monde caribéen*. Le Seuil, Paris.

Ferry, L. (1992). *Le nouvel ordre écologique*. Grasset, Paris.

Fressoz, J.-B. and Bonneuil, C. (2015). *L'Événement Anthropocène*. Le Seuil, Paris.

Gandy, M. (2015). *Écologie Queer. Nature, sexualité et hétérotopies*. Eterotopia, Paris.

GAR (2015). Réduction de risques de catastrophe : bilan mondial [Online]. Available at: www.preventionweb.net/english/hyogo/gar/2015/en/gar-pdf/GAR2015_FR.pdf [Accessed 23 May 2020].

García Acosta, V. and Musset, A. (eds) (2018). *Les catastrophes et l'interdisciplinarité*. L'Harmattan, Paris.

Gautier, D. and Benjaminsen, T.A. (2012). *Environnement, discours et pouvoirs. L'approche political ecology*. Quae, Paris.

Gemenne, F., Rankovic, A., Ansart, T., Martin, B., Mitrano, P., Rio, A. (2019). *Atlas de l'Anthropocène*. Presses de Sciences Po, Paris.

Haraway, D. (2016). Anthropocene, capitalocène, plantationocène, chtulucène. Faire des parents. *Multitudes*, 4(65), 75–81.

Hornborg, A. (2017). Dithering while the planet burns: Anthropologist's approaches to the Anthropocene. *Reviews in Anthropology*, 46(2–3), 61–77.

Hulme, M. (2009). *Why We Disagree About Climate Change: Understanding Controversy, Inaction and Opportunity*. Cambridge University Press, Cambridge.

Keucheyan, R. (2014). *La nature est un champ de bataille. Essai d'écologie politique*. Zone, Paris.

Kull, C. and Batterbury, S. (2017). La place de l'environnement dans les géographies anglophones et francophones. Émergence, transformations et circulations de la *political ecology*. In *Humanités environnementales. Enquêtes et contreenquêtes*, Blanc, G., Demeulenaere, E., Feuerhan, W. (eds). Publications de la Sorbonne, Paris.

Latour, B. (1991). *Nous n'avons jamais été modernes*. La Découverte, Paris.

Latour, B. (2015). *Face à Gaïa*. La Découverte, Paris.

Latour, B. (2017). *Où atterrir ? Comment s'orienter en politique*. La Découverte, Paris.

Lavell, A. and Maskrey, A. (2014). The future of disaster risk management. *Environmental Hazards*, 13(4), 267–280.

Lewis, S. and Maslin, M.A. (2015). Defining the Anthropocene. *Nature*, 519(7542), 171–180.

Lorimer, J. (2017). The Anthroposcene: A guide for the perplexed. *Social Studies of Science*, 47(1), 117–142.

Lövbrand, E., Beck, S., Chilvers, J., Forsyth, T., Hedrén, J., Hulme, M., Lidskog, R., Vasileiadou, E. (2015). Who speaks for the future of Earth? How critical social science can extend the conversation on the Anthropocene. *Global Environmental Change*, 32, 211–218.

Lovelock, J. and Margulis, L. (1974). Atmospheric homeostasis by and for the biosphere: The Gaia hypothesis. *Tellus*, 26(1–2), 2–10.

Malm, A. and Hornborg, A. (2014). The geology of mankind? A critique of the Anthropocene narrative. *The Anthropocene Review*, 1(1), 62–69.

Marsh, G.P. (1865). *Man and Nature; or Physical Geography as Modified by Human Action*. C. Scribner, New York.

Maskrey, A. (ed.). (1993). *Los desastres no son naturales*. La Red, Panama [Online]. Available at: www.desenredando.org/public/libros/1993/ldnsn/LosDesastresNoSonNaturales-1.0.0.pdf [Accessed 8 May 2020].

Mathieu, N. and Jollivet, M. (1989). *Du rural à l'environnement. Penser les changements socio-écologiques*. L'Harmattan, Paris.

Mercer, J. (2012). Knowledge and disaster risk reduction. In *The Routledge Handbook of Hazards and Disaster Risk Reduction*, Wisner, B., Gaillard, J-C., Kelman, I. (eds). Routledge, New York.

Metzger, P. (2017). Connaissances et relations de pouvoir : les sciences sociales, les risques et l'environnement. HDR Thesis, Université Grenoble Alpes, Grenoble.

Metzger, P. and Robert, J. (2015). Environnement et risques : les sciences sociales piégées entre critique radicale et utilité sociale. In *Espaces et rapports de domination*, Clerval, A., Fleury, A., Rebotier, J., Weber, S. (eds). PUR, Rennes.

Moore, J. (2003). Capitalism as world-ecology. Braudel and Marx on environmental history. *Organization and Environment*, 16(4), 431–458.

Moore, J. (2017a). The Capitalocene part I: On the nature and origins of our ecological crisis. *The Journal of Peasant Studies*, 44(3), 594–630.

Moore, J. (2017b). Confronting the popular Anthropocene. Toward an ecology of hope. *New Geographies*, 9, 186–191.

Moore, J. (2017c). The road to ruin? Making sense of the Anthropocene. *IPPR Progressive Review*, 24(3), 176–202.

Moore, J. (2018). The Capitalocene part II: Accumulation by appropriation and the centrality of unpaid work/energy. *The Journal of Peasant Studies*, 45(2), 237–279.

Mumford, L. (1950). *Technique et civilisation.* Le Seuil, Paris.

O'Brien, K., Eriksen, S., Nygaard, L.P., Schjolden, A. (2007). Why different interpretations of vulnerability matter in climate change discourses. *Climate Policy*, 7(1), 73–88.

O'Keefe, P., Westgate, K., Wisner, B. (1976). Taking the naturalness out of natural disasters. *Nature*, 260(5552), 566–567.

Oliver-Smith, A., Alcántara-Ayala, I., Burton, I., Lavell, A. (2017). The social construction of disaster risks: Seeking root causes. *International Journal of Disaster Risk Reduction*, 22, 469–474.

Pahl-Wostl, C. (2009). A conceptual framework for analysing adaptive capacity and multi-level learning processes in resource governance regimes. *Global Environmental Change*, 19(3), 354–365.

Palsson, G., Szerszynski, B., Sörlin, S., Marks, J., Avril, B., Crumley, C., Hackmann, H., Holm, P., Ingram, J., Kirman, A., Pardo Buendía, M., Weehuizen, R. (2013). Reconceptualizing the "Anthropos" in the Anthropocene: Integrating the social sciences and humanities in global environmental change research. *Environmental Science and Policy*, 28, 3–13.

Passeron, J.-C. (1995). L'espace mental de l'enquête (I). *Enquête*, 1, 13–42.

Peet, R. and Watts, M. (1996). *Liberation Ecologies. Environment, Development and Social Movements*. Routledge, London.

Pelletier, P. (1993). *L'imposture écologiste*. Reclus, Montpellier.

Pelling, M. (2010). *Adaptation to Climate Change: From Resilience to Transformation*. Routledge, London.

Pigeon, P. and Rebotier, J. (2016). *Disaster Prevention Policies. A Challenging and Critical Outlook*. ISTE Press, London, and Elsevier, Oxford.

Rebotier, J. (2020). Adaptation. In *Dictionnaire critique de l'anthropocène*, Cynorhodon (ed.). CNRS Édition, Paris.

Rebotier, J., Pigeon, P., Metzger, P. (2019). Returning social context to seismic risk knowledge & management. Lessons learned from an interdisciplinary research in the city of Esmeraldas, Ecuador. *Cybergeo: European Journal of Geography* [Online]. Available at: https://journals.openedition.org/cybergeo/31787 [Accessed 11 May 2020].

Ribot, J. (2014). Cause and response: Vulnerability and climate in the Anthropocene. *Journal of Peasant Studies*, 41(5), 667–705.

Ribot, J. (2019). Social causality of our common climate crisis: Toward a sociodicy for the Anthropocene. In *The Commons in a Glocal World: Global Connections and Local Responses*, Haller, T., Breu, T., De Moor, T., Rohr, C., Znoj, H. (eds). Routledge, London.

Robbins, P. (2004). *Political Ecology: A Critical Introduction*. Blackwell Publishing, Hoboken, NJ.

Robic, M.-C. (ed.) (1992). *Du milieu à l'environnement : pratiques et représentations du rapport homme-nature depuis la Renaissance*. Economica, Paris.

Robic, M.-C. (ed.) (2006). *Couvrir le monde. Un grand XXe siècle de géographie française*. ADFP, Paris.

Rodríguez, H., Quarantelli, E.L., Dynes, R.R. (2007). *Handbook of Disaster Research*. Springer, New York.

Rossi, G. (2001). *L'ingérence écologique*. CNRS Éditions, Paris.

Servigne, P. and Stevens, R. (2015). *Comment tout peut s'effondrer*. Le Seuil, Paris.

Shaw, R. (2013). Disaster risk reduction and community approaches. In *Community Practices for Disaster Risk Reduction in Japan*, Shaw, R. (ed.). Springer, Tokyo.

Smith, N. (2010). *Uneven Development: Nature, Capital, and the Production of Space*, 3rd edition. University of Georgia Press, Athens, London.

Steffen, W., Crutzen, P.J., McNeill, J.R. (2007). The Anthropocene: Are humans now overwhelming the great forces of nature? *Ambio*, 36(8), 614–621.

Steffen, W., Broadgate, W., Deutsch, L., Gaffney, O., Ludwig, C. (2015). The trajectory of the Anthropocene: The Great Acceleration. *The Anthropocene Review*, 2(1), 81–98.

Theys, J. and Fabiani, J.-L. (eds) (1987). *La société vulnérable. Évaluer et maîtriser les risques*. Presses de l'École normale supérieure, Paris.

Tissier, J.-L. (1992). La géographie dans le prisme de l'environnement (1970–1990). In *Du milieu à l'environnement : pratiques et représentations du rapport homme-nature depuis la Renaissance*, Robic, M.-C. (ed.). Economica, Paris.

Tsing, A. (2017). *Le champignon de la fin du monde*. La Découverte, Paris.

White, G.F., Kates, R.W., Burton, I. (2001). Knowing better and losing even more: The use of knowledge in hazards management. *Global Environmental Change Part B: Environmental Hazards*, 3(3–4), 81–92.

Wisner, B. (2016). Vulnerability as model, concept, metric and tool. *Oxford Research Encyclopedia* [Online]. Available at: https://oxfordre.com/naturalhazardscience/view/10.1093/acrefore/9780199389407.001.0001/acrefore-9780199389407-e-25 [Accessed 9 May 2020].

Wisner, B. (2019). Disaster studies at 50: Time to wear bifocals? In *Disaster Research and the Second Environmental Crisis. Assessing the Challenges Ahead*, Kendra, J., Knowles, S.G., Wachtendorf, T. (eds). Springer, Basel.

Wisner, B., Gaillard, J.-C., Kelman, I. (2012). *The Routledge Handbook of Disaster Risk Reduction*. Routledge, New York.

Yusoff, K. (2018). *A Billion Black Anthropocene*. University of Minnesota Press, Minneapolis, MN [Online]. Available at: https://manifold.umn.edu/projects/a-billion-black-anthropocenes-or-none [Accessed 17 May 2020].

PART 1

Toward Unprecedented Risks?

1

Coastal Risks: Coastlines Always Under Pressure

Catherine MEUR-FEREC
LETG, Université de Bretagne Occidentale, Brest, France

1.1. Introduction: environmental risks/natural risks

In a context of climate change and concentration of human activities on the coasts, the risks of coastal erosion and marine submersion are constantly increasing worldwide (Nicholls and Cazenave 2010; GIEC 2013). Thinking about these risks in relation to the Anthropocene means questioning the nature–society interactions that are at their origin. The coastal risks of erosion and submersion are the result of the encounter between intrinsically mobile coasts and static human constructions.

In the midst of the Anthropocene, where environmental concerns occupy a growing place in our societies, coastal risks are often considered (by the Ministry of the Environment, by the natural sciences) as environmental risks or challenges, in the same way as the erosion of biodiversity, for example. However, this term is ambiguous, because it brings together very different problems in terms of the respective places of humans and nature. Is it a danger to the natural environment or a danger from the natural environment? It is conceivable that these are not the same thing, indeed that they are almost the opposite, and the distinction is essential when it comes to managing these risks. The ways of posing the problems, of proposing solutions and of arguing them are totally different: defending nature or defending against nature? Moreover, the terms used in management policies reflect this ambiguity. If the actions of "coastal protection" or "coastal defense" seem to be a consensus in our societies, for some they mean protecting the "wild" coasts against

Risks and the Anthropocene,
coordinated by Julien REBOTIER. © ISTE Ltd 2021.

excessive urbanization and associated pollution, and for others protecting the inhabitants and their property against attacks from the sea.

Basically, and more generally, the major difference between these two positions is the place of the human in the problem of "environmental" risks. Either (i) he is perceived as a disrupter, a waster of resources, putting humanity in danger by his excesses: then we are criticizing excessive and harmful development for the planet. It is our planet as our home that is at stake, and man who is the risk. Management then consists of regulating human actions to slow down the degradation caused to nature. Or (ii) it is man who is perceived as threatened by natural hazards (floods, volcanic eruptions, tsunamis, etc.). The man represents the stakes, when the human lives risk being lost and when the goods (houses, roads, networks, etc.) risk being destroyed. Management in this case is to try to control destructive natural forces to protect human stakes and prevent disasters (Pigeon *et al.* 2018). When it comes to risk, defining what is considered a hazard and what is considered a stake appears to be an essential prerequisite to any analysis.

However, the Anthropocene moment implies a paradigm shift that tends to take place in terms of risks related to nature. The vision of devastating nature finds less echo in the approach of scientists and societies (particularly the wealthy ones): nature is more and more presented as a stake to be defended against human destruction rather than a dangerous hazard. On the subject of coastal risks, this paradigm shift is particularly notable and places these risks in an ambiguous position. While they are considered as "environmental challenges" that place nature at stake in many research programs and political discourses, they are understood as "natural risks", the danger of which is caused by nature, in most management measures, such as in the Natural Risk Prevention Plans in France (*Plans de prévention des risques naturels*, PPRN).

This distinction between nature and society may seem excessively dichotomous. Geography is precisely the discipline that studies the relations between these two theoretical entities through systemic approaches where all the elements are in interrelations within the ecumene (Berque 1996). In the field, everything communicates through porous boundaries, as evidenced by the well-known anthropization of many natural hazards (D'Ercole and Pigeon 1999). As far as coastal risks are concerned, if it is admitted that the hazards are attributable to natural dynamics (geomorphological and hydrodynamic processes generated by meteorological and marine forcing), it is also well known that these natural hazards are often aggravated by anthropogenic factors (sediment extraction, harbor structures, dikes, etc.). Furthermore, the distinction between natural and anthropogenic factors is made even more difficult by the influence of climate change on coastal hazards, the human component of which is now demonstrated. It is therefore not easy to distinguish between natural and anthropogenic components in coastal hazards. However, two things are clear: the main stakes in this problem are human (human lives and built stakes), and in terms of management, the objective is to avoid damage to these stakes as much as possible. The coastal risks are therefore for us "natural risks". This precision is essential to conduct a scientific analysis and even more so to draw lessons for society.

Section 1.2 presents the importance of coastal risks in the world during the 20th century and the links with climate change. Section 1.3 discusses the concept of systemic vulnerability of coastal territories and its application to the French coastline. Section 1.4 takes a reflective look at the challenges of action research and what the notion of the Anthropocene can contribute.

1.2. Desire for shores and climate change: the increase in coastal risks in the world in the 20th century

As with all so-called natural disasters, coastal hazards have taken on an unprecedented scale throughout the world in the 20th century. It is often difficult to identify them precisely in international statistics because they are partly classified as *floods* and *storms*. In France, they are also listed under landslides for erosion and cliff failures. Even if there is no worldwide data on coastal damage, their growth follows that of other natural disasters. As for any risk, the causes of the increase in damage are to be found in the intersection of hazards and stakes. While the former show a tendency to worsen, which is highly echoed by the media in a context of climate change, it is the latter that have multiplied most significantly.

1.2.1. *Coastal hazards and climate change: the interweaving of several spatial and temporal scales*

The strong storms, cyclones and hurricanes of the last decades and their consequences on the coasts on a global scale have led the media, and sometimes scientists, to emphasize the direct causal links that may exist between climate change and the current evolution of shorelines subject to erosion and submersion.

To understand this link, which is less obvious than it seems, it is necessary to put contemporary coastal geomorphological dynamics into perspective in the context of their evolution (Hénaff *et al.* 2013). They are part of processes acting on different intertwined temporal scales: the postglacial transgression at the origin of the establishment of the current coastlines, the historical evolutions of the shorelines and the permanent dynamics of the coastal forms subjected to meteorological and marine phenomena.

First of all, the development of the present coasts is largely the result of geological time scale conditions and climatic fluctuations of the quaternary period. The last cold period and the marine transgression that followed it created massive accumulations that constitute our current beaches, pebble strips, coastal dunes and maritime marshes. These inherited sedimentary stocks are no longer massively fed today, but they are permanently reworked by an intense geomorphological dynamic.

Then, over a secular or multi-decadal time span, the lasting evolutionary trends of a stretch of coastline respond to the actions of natural morphogens (variations in storm intensity and frequency, changes in wind and swell orientation, modification of continental hydrology, etc.) or increasingly anthropogenic ones as we approach the contemporary era (harbor structures, sediment extraction, sea defense structures, etc.).

Finally, on the scale of meteorological and marine events, significant changes in coastal forms can be observed during storms, especially if they are coupled with certain tidal conditions on the meso- and macrotidal coasts (high tidal coefficients and high tide). Thus, strong swells, heightened by marine surges often exceeding 1 m related to wind and low atmospheric pressure, can directly attack the tops of beaches and cause, in a few hours, spectacular erosion phenomena: retreat of the coastline of several tens of meters, lowering of beaches, breaches, submersions. Moreover, a succession of stormy events weakens the coasts all the more as the sedimentary compensations that occur naturally after each event do not have time to take place before the arrival of the next storm. These paroxysmal events, which are brutal, generate considerable instantaneous coastal recession rates, but they do not constitute a trend allowing reliable predictions, which only a long period of observation allows.

In this dynamic, the current annual millimeter sea level rise related to climate change, even if it is increasing, is very small compared to the metric height of storm surges. Sea level satellites – TOPEX/POSEIDON, then Jason-1 – show an annual rise of 3 mm since the early 1990s (Cazenave and Llovel 2010). In comparison, the surge recorded at La Rochelle during the Xynthia storm in 2010 was 1.53 m and the quite exceptional surge of the storm of February 1, 1953, in the southwestern Netherlands was 3 m. Thus, in current erosion and flooding events, the eustatic signal related to climate change is weak, masked by one-off events of much greater amplitude. However, the sea level rise predicted by the IPCC (Intergovernmental Panel on Climate Change) for the year 2100, which was estimated in 2013 to be between 26 cm and 98 cm depending on the predicted CO_2 emissions and the models used, and revised upwards in 2018, cannot be neglected (GIEC 2013, 2018).

Furthermore, it is logical to think that global warming influences extreme weather events, which is already clearly evident in the case of tropical cyclones. In temperate zones, however, where strong regional disparities are apparent, results on this subject are not yet stabilized. In any case, and whatever the intensity of the eustatic rise, it will raise the instantaneous water levels of storm events, increasing the risk of erosion and submersion.

Two other major reasons, directly linked to human activities, aggravate this already unfavorable situation in terms of coastal hazards. On the one hand, the stocks inherited from the cold periods of the quaternary have been diminished by a number of coastal extractions and are now threatened by the growing demand for marine aggregates for construction (ONU 2019). On the other hand, the multiplication of static works to defend built-up areas against the sea freezes the coastline and blocks dune-beach exchanges. It induces effects of thinning beaches and postpones, by exacerbating them, the erosive processes on other sectors (Ministère de l'Écologie 2010).

Global warming thus appears to be a factor that aggravates coastal hazards, but at the moment it is not the driving force in a context that is already very unfavorable to coastal stability.

In addition to its scientific inaccuracy, this amalgam between meteorology and climate (enshrined in the widespread, albeit antinomic, expression "climatic event"), directly linking coastal dynamics and climate change, has a twofold effect on coastal use. On the one hand, it contributes to removing responsibility from the inhabitants and decision-makers behind the urbanization of exposed sectors. This is because climate change is a relatively new paradigm, which was little talked about half a century ago, and because it is a global phenomenon, which tends to be spatially and temporally distant (see section 1.3.2.3). On the other hand, climate change is also an essential tool for awareness raising and prevention, as its mere mention is enough to remind citizens and elected officials that the coast was, is and will remain a territory at risk. It gives more resonance to speeches aimed at limiting the stakes in the coast, and even at considering relocating these stakes toward the land in the perspective of a sustainable development of coastal territories.

However, like all risks, the driving force of coastal risks is not limited to hazards. It is multiple, and also concerns the stakes which tend very strongly, since the 20th century and on a global scale, to multiply near the coasts.

1.2.2. *Coastalization of population and activities, the main cause of coastal risks in the world*

The phenomenon of coastal settlement and activities, long analyzed by geographers such as D. Noin (1999), has been exacerbated since the second half of the 20th century. The trend is toward a strengthening of the phenomenon, particularly in the low-lying coastal areas of developing countries (Neumann *et al.* 2015).

According to the global synthesis by Neumann *et al.* (2015), five Asian countries (China, India, Bangladesh, Indonesia and Vietnam) accounted for more than half of the world's low-lying coastal population in 2000 and will continue to do so in the future, despite the rapid population growth of several African coastal countries (Figure 1.1). Most of the world's megacities of more than 8 million people are located in the coastal zone, and many are developing in deltaic areas where specific economic, geographic and historical factors combine to drive migration and increase the coastlization of settlement. In China, the growth of coastal urban areas, associated with economic development, is particularly high, more than three times the national rate.

For thousands of years, the world's coastal zones have attracted populations for various reasons depending on the territory and the time. Agricultural, fishing, military, industrial, commercial, leisure or residential reasons have led to the development of activities and the installation of societies near the coastline. This trend has been reinforced for several decades with the development of coastal urbanization and the ever-growing attraction of our contemporary societies to the seashore.

There are several reasons for settling in an area exposed to coastal hazards. For a large part of the world's population, the primary motivation is related to work near the sea. Whether one is a fisherman, a port worker, or a tourism employee, the land–sea interface constitutes a work space. These populations, often socially vulnerable, reside close to their place of work, and therefore frequently in areas exposed to coastal risks. For other populations, the coasts can be places of relegation, for migrants from the sea or for people illegally occupying the public domain, for example, the "fifty steps" zone, in French overseas territories. They are illegal occupants, without property title and in the front line of coastal hazards that are particularly formidable in intertropical climate. Finally, other coastal inhabitants are, on the contrary, privileged populations of the most developed countries who voluntarily seek the proximity of the sea to take advantage of its amenities: a view, direct access to the beach, the beauty of the landscape. This is the case in metropolitan France, for example.

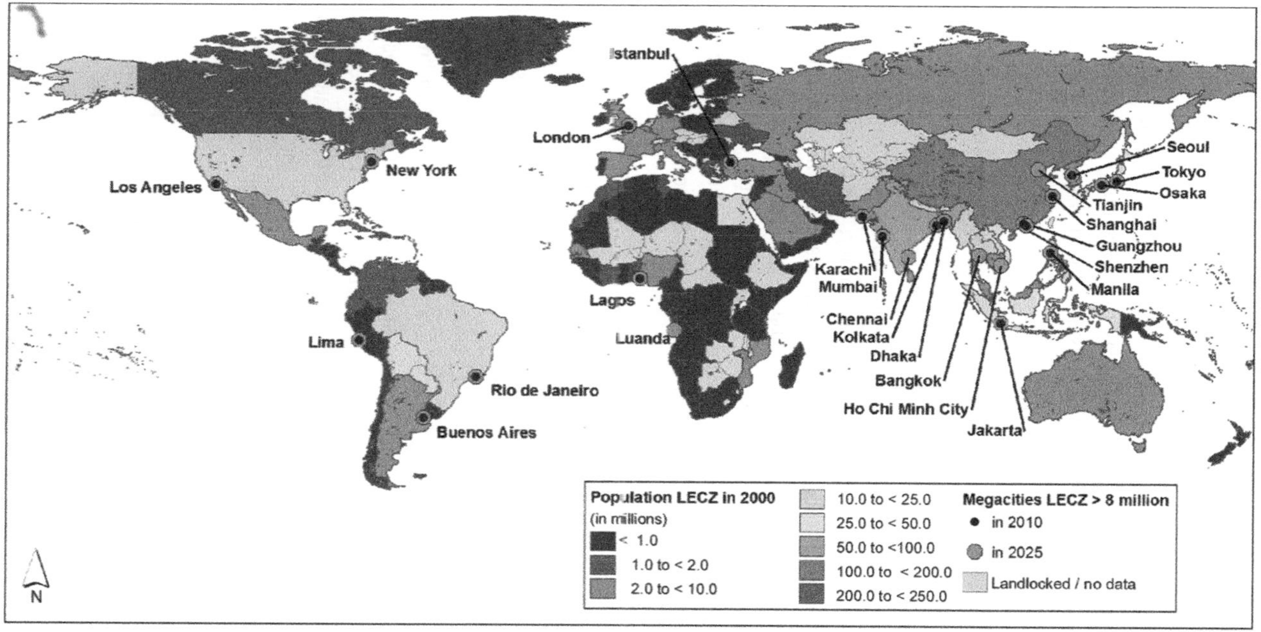

Figure 1.1. *Map of the world population in 2000 and the estimated population in 2010 and the projected population in 2025 of a selection of megacities located in low elevation coastal zones (LECZ), according to the UN's world urbanization forecasts. Although the rules of graphic semiology are not followed on this map (population quantity represented by colored solids over the entire surface of the countries), we found it informative enough to keep (source: from Neumann* et al. *2015). For a color version of this figure, see www.iste.co.uk/rebotier/risks.zip*

1.2.3. *Coastal risks caused by the desire for shores*

The work of historians reveals that the densification of the occupation of the coasts of Western countries is relatively recent. Until the end of the 18th century, most European coastlines were sparsely occupied (Corbin 1988). The shores did not attract crowds and the occupation of the coastline responded to the needs of activities requiring proximity to the sea. It was thus essentially utilitarian reasons, those of the working coast, that motivated coastal settlements, and not an attraction for a living or leisure environment.

Progressively, representations of the coastline have changed. In parallel with the democratization of society and the transition from a rural to an urban society, the rise of seaside tourism has radically changed the relationship of man to the shoreline. While the seaside became a place for walks and invigorating baths recommended by doctors at the end of the 18th century, the coastal area began to constitute a spatial resource to be exploited for real estate. During the 20th century, the few aristocratic villas built along the coast gradually gave way to more numerous constructions, housing estates and buildings. After the Second World War, many of the old villages that were located a few kilometers from the coastline were duplicated by the creation of seaside resorts located on the shore (Toulier 2016):

> The marine dream creates wealth and new forms of conquest of coastal areas. The sandy shores become coveted beaches fixing new seaside cities with only ludic functions, the seaside resorts. (Cabantous *et al.* 2005)

Thus, the rise of seaside tourism, first aristocratic in Europe and then "mass" and globalized, has contributed to the densification of the occupation of the coast and to bringing buildings ever closer to the shore (Figure 1.2). In parallel with the development of tourism, the new image of the coastline, healthy, natural and fun, reinforces a strong residential tropism. This is expressed in the stretching out of residences (primary or secondary) along the shores, from urban, port or seaside centers.

In this context, the coastal risks of erosion and submersion are particular. They do not result from involuntary exposure to a hazard, like most natural and technological risks, but from a desired proximity to a sea that attracts and makes people dream. However, this sea is ambivalent, carrying amenities that are highly prized by socially privileged and mostly urban populations, but also the engine of coastal hazards. The question of social representations of the sea, which change according to place and time, therefore appears to be essential to understanding the mechanisms that generate stakes, and therefore coastal risks (see section 1.3.2.3).

Figure 1.2. *Blankenberge beach, Belgium (source: C. Meur-Ferec, September 2005). For a color version of this figure, see www.iste.co.uk/rebotier/risks.zip*

When one crosses this trend of voluntary movement of stakes toward the sea, with that of the retreat of the shores toward the land in a context of sedimentary shortage and rising sea level, one understands the emergence and multiplication of coastal risks, in societies privileged to the Western way of life, during the 20th century. These converging dynamics can be found on most of the world's urbanized seafronts (Figure 1.3).

In this diagram, the "buffer zone" of a few kilometers between the old villages and the coastline was gradually reduced between the end of the 19th century and the beginning of the 20th century. The disappearance of this "safety area" was largely accelerated after the Second World War, during the seaside boom of "mass" tourism, from which the constructions were brought as close as possible to the coastline in order to take maximum advantage of all the amenities offered by the sea. It is in this situation of predictable telescoping between hazards and stakes that the risks of erosion and submersion have emerged. Faced with the appearance of these risks, the reaction of the owners of property exposed to the mobility of the coastline was to defend them by trying to fix the shoreline, against a natural tendency to regress. Danger spaces have then developed and are maintained permanently, in an artificial way, in a relative and very expensive stability. Climate change and the consequent acceleration of sea level rise clearly appear as factors aggravating the hazards, presaging ever-increasing damage to urbanized coastlines.

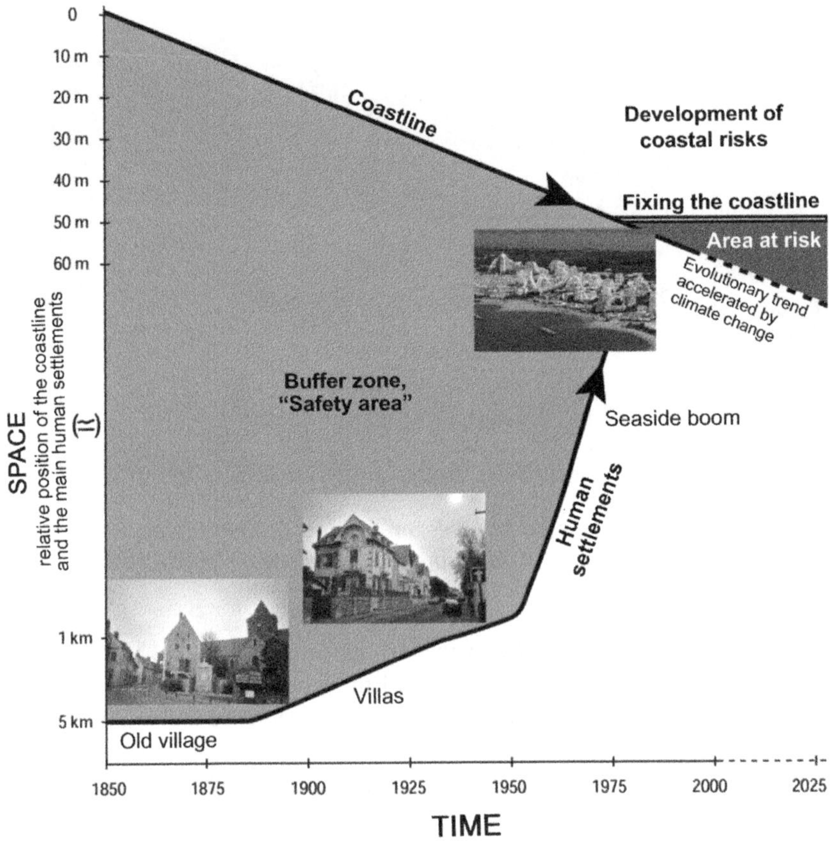

Figure 1.3. *Diagram of the emergence and multiplication of coastal risks in the 20th century on the French seaside (source: adapted from Meur-Ferec and Morel 2004). For a color version of this figure, see www.iste.co.uk/rebotier/risks.zip*

Analyzing the evolution of hazards and stakes allows us to estimate the evolution of coastal risks, but other dimensions are missing to understand the dynamics and to understand in depth the vulnerability of coastal territories.

1.3. Systemic approach to the vulnerability of coastal territories

Vulnerability is a polysemous term (Birkmann 2006) that offers the possibility of taking a systemic and integrated approach to coastal risks (Meur-Ferec *et al.* 2020). When considered solely as the social component of risk, this concept does not provide the services that one would expect from it in terms of knowledge synthetics and decision support. The concept of "systemic vulnerability" proposed in this section integrates all the factors (natural, economic, political, social) of fragility of a coastal territory. Vulnerability is understood as a system with multiple interacting components.

1.3.1. *For an integrated and broadened conception of vulnerability*

The scientific literature on the subject of vulnerability is abundant and shows first of all that a consensus is far from being established on its meaning and use (Gilbert 2009). The use of this notion has also developed strongly over the last 20 years in publications concerning the impacts of climate change. Even in the limited field of coastal vulnerability, the diversity of uses of this term remains very large.

Vulnerability is specific to each territory, intimately linked to its history, its use and its population. In this sense, it has an eminently geographical character, as it is strongly place based (Cutter *et al.* 2003; Rebotier 2012). Also, because it is socially situated, vulnerability cannot be studied in a universal way. This is all the more true when we give an important place in its analysis to questions of governance and representations, which are particularly sensitive to social, political or administrative contexts.

Among this proliferation of uses of the notion of vulnerability, we have progressively developed the concept of "systemic vulnerability" to coastal flooding-erosion risks through various research projects (Meur-Ferec *et al.* 2008; Hellequin *et al.* 2013; Hénaff and Philippe 2014). This approach of course builds on previous work, in particular that of Robert D'Ercole (1994), which defines the vulnerability of societies through their capacity to respond to potential crises, a capacity that depends on both incidental (the hazard) and structural (social, economic, cultural, functional, institutional context) factors. One of the particularities of this approach is therefore to consider hazards as an integral part of vulnerability, whereas they are generally studied separately. Integrating hazards into vulnerability makes it possible to avoid "a manicheistic and naive reading of the hazard-vulnerability pair (opposing nature on the one hand and culture on the other)" (D'Ercole and Pigeon 2000).

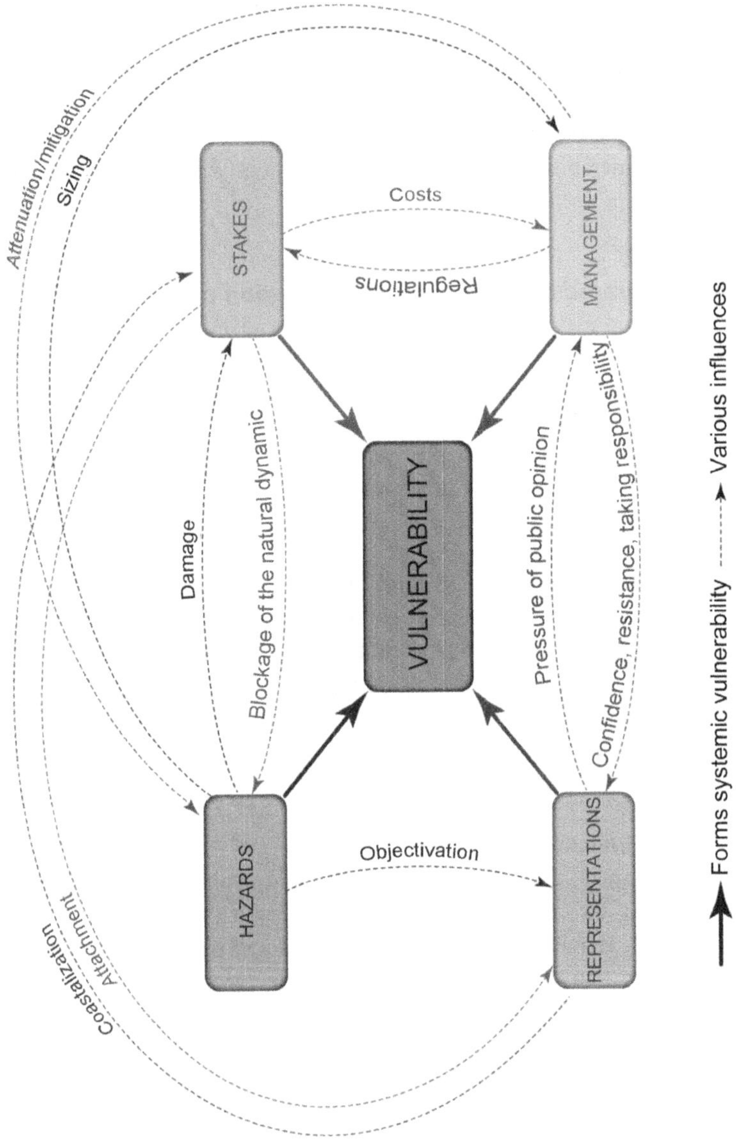

Figure 1.4. *Systemic vulnerability diagram (source: from Meur-Ferec et al. 2008). For a color version of this figure, see www.iste.co.uk/rebotier/risks.zip*

In this sense, vulnerability is no longer just the social parameter of risk, but becomes a resultant, expressing the fragility of a territory as a whole. Vulnerability as a system is thus the result of the combination of four interdependent components (Figure 1.4). Classically, hazards (1) (in this case coastal erosion and marine submersion) are processes of more or less natural origin likely to damage or destroy the stakes (2) that are exposed to them. The stakes are the people, the goods and the activities that are located in a territory exposed to the hazard. These two components make it possible to define the risk, but are insufficient to assess vulnerability. Two other components are therefore taken into account. Risk management (3) includes public policies for protection, prevention and crisis management and their application by risk management actors in the field. And representations (4) take into account the relationship to risk of the populations present in the territory concerned and of the actors responsible for its management (sensitivity to risk, relationship to place, adaptation preferences, etc.).

These four components combine to form the systemic vulnerability of an area at a given time (Meur-Ferec *et al.* 2008). It is possible to enter the system through any of the components; the important thing is to study the four and take into account their respective contributions to the constitution of vulnerability (Figure 1.4, solid arrows). First of all, hazards and stakes are essential components, because they constitute risk: without either of them there would be no risk, and therefore little interest in assessing management and representations. By the same token, these two components have a decisive influence on the overall diagnosis of vulnerability: if the hazard is low and the stakes are few, a lesser sensitivity of the population to the risk and a management that takes little account of it are of much less importance than if the risk were high. Hazards and stakes clearly increase vulnerability. Management, on the other hand, is in principle conducive to reducing vulnerability (although inappropriate management can have negative effects). The influence of representations on vulnerability is much more difficult to grasp. It can contribute to increasing or decreasing it. Finally, the four components are largely interdependent (Figure 1.4, dotted arrows). For example, management influences the stakes by regulating construction in exposed areas; representations of the coast as a privileged place to live influence the stakes by increasing the coastalization of the population; hazards influence management by conditioning the choice and dimensioning of protective structures, etc.

The systemic vulnerability approach therefore involves a multi-criteria analysis. It also makes it possible to identify the components that contribute most to overall vulnerability. Depending on the area, this may be due to high hazards, a high concentration of stakes on the coastal fringe, a lack of consideration of risk in urban planning documents and/or a lack of interest on the part of inhabitants.

1.3.2. *The systemic vulnerability of coastal areas in France*

Analysis of systemic vulnerability is difficult to consider at the global level. However, it is possible to outline the major trends in the four components of systemic vulnerability at the scale of metropolitan France.

The current French context has been strongly marked by the dramatic event of the Xynthia storm in 2010, which caused 47 deaths on the Atlantic coast and placed coastal risks at the forefront of media and political scenes. It revealed the extent of the phenomenon and the complexity of the problem in a country that has relied on the residential and tourist development of its coasts.

1.3.2.1. *Hazards: part of the problem*

Metropolitan France is sheltered from tropical cyclones; it is not subject to major tsunamis or strong subsidence; its varied topography does not predispose it particularly to submersion; and the lithological diversity of its coasts gives it variable erosion speeds. Compared to other regions of the world, such as Southeast Asia or the Caribbean, for example, it is not particularly exposed to coastal hazards, even if some parts of its territory are very sensitive, especially in the overseas territories. However, many studies are devoted to coastal hazards. Within the framework of PPRNs, their mapping is the subject of sophisticated hydrodynamic modeling and gives rise to passionate debates on water heights and recession speeds (Perherin *et al.* 2017). Numerous coastline observatories are organized in a network by the Ministry of the Environment, which commissioned the mapping of low-lying coastal areas and the national erosion index following Xynthia (Figure 1.5).

Coastal hazards are often aggravated by human actions. Extraction of sand and pebbles is now strictly regulated on land and in the shallows, but the sedimentary stock has been heavily exploited since the Second World War, mainly by the construction industry. For example, the pebble strip in Audierne Bay, which was mined for decades to make concrete, has lost more than half its pre-war volume (Figure 1.6).

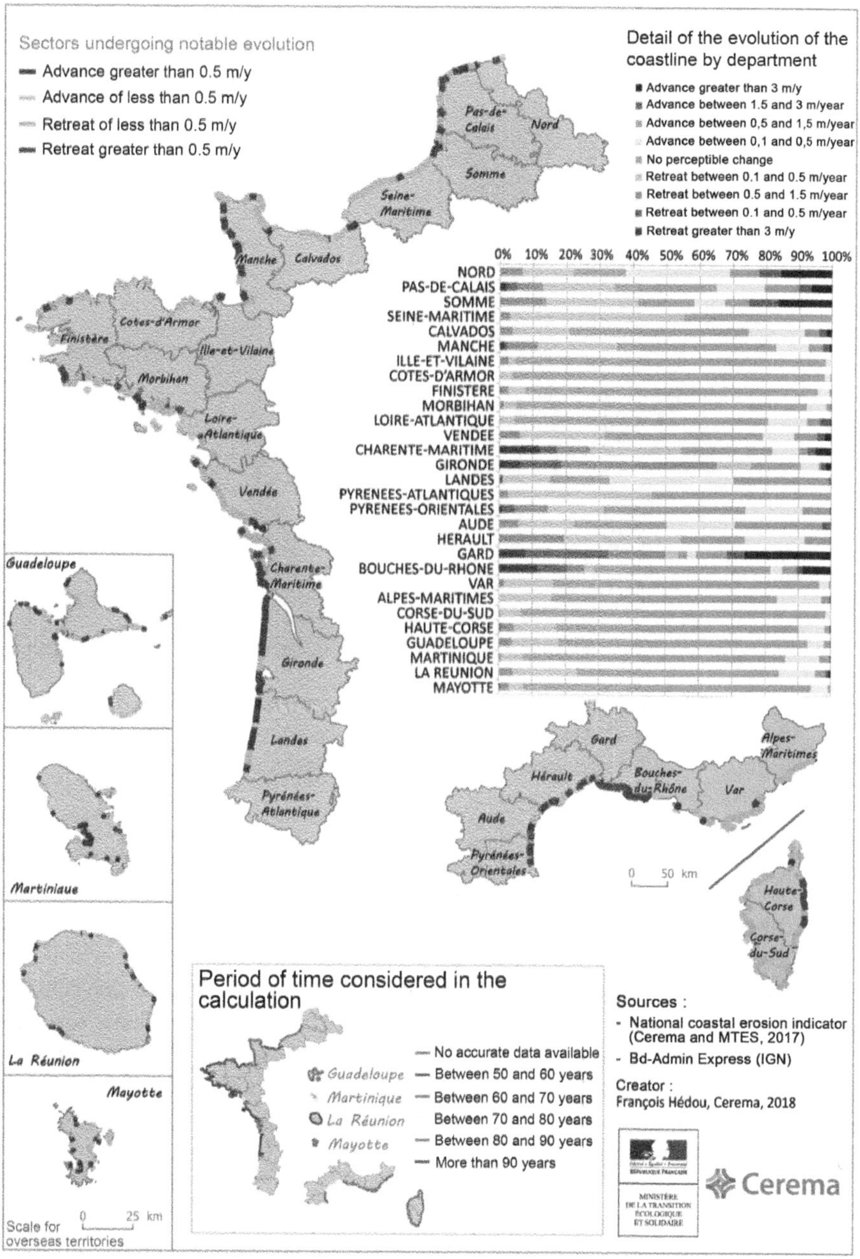

Figure 1.5. *Map of the national coastal erosion indicator in France (source: Cerama 2018). For a color version of this figure, see www.iste.co.uk/rebotier/risks.zip*

Figure 1.6. *Remains of the pebble crushing plant of Tréguennec, Audierne Bay, France (source: C. Meur-Ferec, 1989). For a color version of this figure, see www.iste.co.uk/rebotier/risks.zip*

In addition, according to a cartographic inventory carried out by the Center for Studies and Expertise on Risks, Environment, Mobility and Development (Cerema 2018), about 16,000 structures and developments are present on the French coastline, over a cumulative length of 2,300 km. These structures, which have increased significantly since the 1960s, artificialize about 30% of the coastline. While their purpose is to fix the shoreline in high-stake areas, they have secondary effects on the erosion of beaches and adjacent areas (Figure 1.7). They contribute both to reducing the hazards of submersion and erosion in a localized manner, and to increasing them in the long term by hindering the natural mobility of the coasts and reducing their capacity for resilience.

As in other coastal areas of the world, the trend toward erosion and submersion of the French coastline is expected to continue or to increase in a context of global warming and a largely depleted sedimentary stock.

Figure 1.7. *Lowering of the beach at the foot of a structure, Wissant, France (source: C. Meur-Ferec, 2003). For a color version of this figure, see www.iste.co.uk/rebotier/risks.zip*

1.3.2.2. *The stakes: the heart of the problem*

In terms of human lives at risk, France is not among the territories with the highest stakes in the world. Unlike countries such as Bangladesh and Indonesia, for example, the majority of French coastal populations are well housed, informed and financially well off. On the other hand, in terms of exposed property, the French coastline concentrates important built-up areas, the construction of which has been stimulated by a national spatial planning policy which, since the 1960s, has relied heavily on the development of seaside tourism (planning missions for the Languedoc-Roussillon coast in 1963 and the Aquitaine coast in 1968). On other coasts, such as Brittany or the Côte d'Azur, construction developed more spontaneously, but also exponentially in the second half of the 20th century.

In addition to their number, the value of homes exposed to coastal hazards reaches peaks[1] that do not weaken despite damaging events, and even dramatic episodes like Xynthia. The average sale price of building land in coastal municipalities was 60% higher than the average for metropolitan France in 2012 and has continued to rise, increasing by 40% between 2006 and 2012 (SOeS 2013).

1 Law no. 86-2 of January 3, 1986 relative to the development, protection and enhancement of the coastline.

Nothing in the real estate markets seems to indicate a reversal of this trend[2]. This point constitutes one of the nodes of the problem of coastal risk management in France, as such real estate values do not allow for public land control of the most exposed properties, as foreseen in the national strategy for integrated coastline management (Ministère de l'Écologie 2012).

This coastalization of the population, and especially of constructions that consume a lot of space, is continuing today, even if the Coastal Law of 1986[3] and the Barnier law of 1995[4] (establishing the PPRNs) have regulated the large amounts of coastal urbanization taking place. Today, coastal municipalities, which represent 4% of the surface area of mainland France, concentrate more than 10% of the population (which can be multiplied by 10 in the summer season), 12% of new housing surfaces, 23% of second homes and 37% of the country's tourist accommodation capacity (Ministère de la Transition écologique 2017). Thus, coastal issues have multiplied closer to the sea for more than half a century. As the craze for seaside residences grew, construction moved closer to areas exposed to coastal hazards.

In addition to their number, the value of homes exposed to coastal hazards reaches peaks[5] that do not weaken despite damaging events, and even dramatic episodes like Xynthia. The average sale price of building land in coastal municipalities was 60% higher than the average for metropolitan France in 2012 and has continued to rise, increasing by 40% between 2006 and 2012 (SOeS 2013). Nothing in the real estate markets seems to indicate a reversal of this trend[6]. This point constitutes one of the nodes of the problem of coastal risk management in France, as such real estate values do not allow for public land control of the most exposed properties, as foreseen in the national strategy for integrated coastline management (Ministère de l'Écologie 2012).

The understanding of such a craze for seaside residence, in a context of coastal risks exacerbated by climate change, is to be sought in representations of the sea and of coastal residence.

---

2 Law no. 95-101 of February 2, 1995 relating to the reinforcement of the protection of the environment.

3 Law no. 86-2 of January 3, 1986 relative to the development, protection and enhancement of the coastline.

4 Law no. 95-101 of February 2, 1995 relating to the reinforcement of the protection of the environment.

5 Eugénie Cazaux's thesis in progress: Effects of the ambivalence of the sea on the land and property values of the French coastline: coastal risks and climate change versus desire for the shoreline. LETG Laboratory, University of Brest.

6 As shown in an article titled "Real estate prices are rising everywhere on the French coast" in *Le Monde* on August 2, 2018 in the Money & Investments section.

1.3.2.3. *Representations: a main component often forgotten*

The “representations” component has long been neglected in risk studies, but its importance was dramatically highlighted in France by the Xynthia storm disaster (Hellequin *et al.* 2013). Several studies on this topic have analyzed the results of surveys conducted among residents of communes exposed to coastal risks, on different coasts of France (Hauts-de-France, Occitanie, Brittany) (Michel-Guillou and Meur-Ferec 2017; Meur-Ferec and Guillou 2020). Among the main results, we note that:

– contrary to certain initial hypotheses, the inhabitants interviewed are not ignorant of coastal erosion and submersion processes. However, the vast majority of them do not spontaneously think of these phenomena when risks are mentioned in their community. They are not worried because they do not “objectify” them as risks (Moscovici 2001). “A big gust of wind” or “a little water in the house” are elements they have chosen to live with, without thinking about them daily in terms of risk;

– coastal risks related to climate change are known but little mentioned at local level. Inhabitants situate these risks outside their place of residence (“it’s in Bangladesh”), or on a temporal scale beyond their lifetime (“in 2100, I won’t be here anymore”). Distancing oneself from the risk helps to maintain a sense of well-being. It distances the dangerous phenomenon by making it an abstraction (Trope and Liberman 2010);

– erosion and flooding are not perceived as insurmountable problems. The inhabitants concerned feel that they are manageable through prevention and protection measures. The knowledge of the environment, combined with this sense of control, allows individuals to maintain the positive identity of their place of living;

– living near the sea is considered above all as an advantage, and even as a privilege. We find here the attraction for the coastline and its positive image, characteristic of contemporary Western societies, which largely conceals the dangers of proximity to the sea;

– in terms of development, the preference for hard structures, aimed at fixing the coastline to protect property, can be linked to the identity dimension of attachment to a place. Thus, coping strategies are accepted as long as they do not lead to changes that result in too much modification of living spaces or relocation, and thus do not interfere with the identity of the place of habitation (Twigger-Ross and Uzzell 1996). Otherwise, a process of resistance tends to lead people to generally endorse national or international strategies, such as relocation, while pointing out the impossibility of their local realization. In addition, political factors, such as strong public intervention to repair damage and compensate property owners, also help explain people's sense of control and their willingness to "hold the line".

Knowledge of representations is an essential element of knowing the systemic vulnerability of a territory. Not only does it allow for a better understanding of public policy orientations and management choices, but also, through the representations of the inhabitants, to explain certain difficulties in implementing national strategies at the local level.

1.3.2.4. *Management: the lever of public policies*

In France, the principles of the welfare state remain strong, compared to other countries that are more liberal economically and socially, such as the United States or Canada. Thus, the management and induced costs of coastal risks are traditionally borne by the public authorities, which regulate urbanization and finance most sea defense works (Meur-Ferec and Rabuteau 2014). In addition, victims of "natural disasters" benefit from a public compensation system based on national[7] solidarity (Cazaux *et al.* 2019). Today, however, in a context of scarce public funds and rising sea levels, and after the Xynthia disaster in 2010, coastal risk management is undergoing significant changes.

First, the State has accelerated measures to control urbanization in coastal areas by identifying 58 priority coastal PPRNs in 2011. Their implementation, involving strict restrictions on building authorization in hazard zones, revealed the often conflicting debates between the State, accused of being too secure, and local authorities, suspected of prioritizing the development of their territory and the defense of their constituents' interests to the detriment of the general interest (Perherin *et al.* 2017). In addition, in 2012, the government implemented the National Strategy for Integrated Coastline Management advocating the relocation of stakes and activities most at risk (Ministère de l'Écologie 2012; Mineo-Kleiner and Meur-Ferec 2016). A governmental reflection is underway on the feasibility of a "spatial recomposition of coastal territories" in a context of climate change (Buchou 2019; CGEDD 2019). If the State encourages the destruction and the retreat of certain stakes, almost no actor on the ground concretely considers this option. The choice of withdrawal, in addition to the immense financial problems it poses in terms of real estate values on the coasts, clashes with social representations of a dreamed coastline. However, it is to be hoped that the specter of climate change and the highly publicized "rising sea level" will help to move things forward in this perspective.

---

7 This system, which is not well adapted to coastal risks, covers flooding and landslides on rocky cliffs, but not dune erosion.

At the same time, and in an often disconnected and even contradictory way, the National Strategy for Flood Risk Management (2014) recommends that local authorities set up action plans for flood prevention (Papi), which opens up state funding for them to create or strengthen sea defenses. Thus, territories that are subject to both erosion and flooding hazards (which is very common on soft coasts where they are intrinsically linked) have every interest in adopting the strategy of flood risk management rather than coastline management if they want to strengthen their position near the sea. Only territories exposed to erosion risks alone are forced to consider long-term adaptation to climate change.

Another significant fact in terms of management is that the welfare state is gradually disengaging, as evidenced by the recent transfer of the "management of aquatic environments and flood prevention"[8] competency to municipalities and their groupings since 2018. The "flood prevention" competency concerns the control and financing of works against river and marine flooding and is therefore a central competency in terms of coastal risks. This decentralization of the management of sea walls constitutes a major upheaval in the current governance of coastal risk management in France. On the one hand, we can hope that it will lead to a greater responsibility of coastal territories and inhabitants. On the other hand, we may fear that it will allow some, the richest, to free themselves from the principles of sustainable development and adaptation to climate change and that it will accentuate inequalities between coastal territories. These considerations on coastal risk management underline that it is less a technical problem than a political one. It raises questions of land use planning, social choices, adaptation to a changing environment and social equity.

In the end, if the hazards and human stakes give France a moderate level of risk on a global scale, the vulnerability of the French coastline is mainly due to two components: on the one hand, public policies that struggle to make inhabitants responsible, without sacrificing the principle of national solidarity, and, on the other hand, a low representation of coastal territories as areas at risk. These two elements, linked together, constitute powerful brakes on the challenge of adaptation posed by climate change.

Thus, the systemic vulnerability approach makes it possible to integrate all the dimensions necessary for understanding territories subject to coastal risks. It underlines the need to go beyond hazard-centered approaches, but also to go beyond purely sociocentered research that ignores natural dynamics. It is in the systemic approach, dear to geographers, that the vulnerability of coastal eco-socio-systems

8 Maptam law of January 27, 2014 on the modernization of territorial public action and the affirmation of metropolises.

can be understood in all its complexity. This understanding can provide enlightening elements for managers in the field (local and state government departments, elected officials), at this Anthropocene time.

1.4. Interests and limits of the Anthropocene moment for thinking about coastal risks

The integrated approach of systemic vulnerability underlines the complexity and multi-causality of coastal risks as well as the importance of their territorial contextualization. Nowadays, the notion of the Anthropocene is regularly mobilized to give a framework and meaning to the "global" issues of the moment. In this last section, and as a reflexive look at the research experience, we will start from some key elements specific to this notion to estimate its interest and limits in the analysis and management of coastal risks.

1.4.1. *The world as a place*

The Anthropocene is thought of above all on a planetary scale. While this scale can be meaningful for defending major causes such as the protection of resources and the natural environment, it quickly reaches its limits. It is impossible to understand a territory's mechanisms of vulnerability without anchoring the analysis to the scale of eco-socio-systems, which are necessarily restricted to present a minimum of homogeneity. The analysis of risks implies a necessary territorialization, in the sense of being anchored in the field, which is dear to geographers (November 2002; Rebotier 2012).

1.4.2. *Man, an actor with a strong influence on the environment*

The notion of the Anthropocene emphasizes the important place of humans (and their responsibility) in major environmental issues. Integrating the human dimension into understanding of the environment and risks is by definition a matter of course for geographers who study nature–society relations. The relatively new Anthropocene approach does not therefore revolutionize our research on natural risks, which is undoubtedly a social construct (Pigeon *et al.* 2018). The Anthropocene is perhaps more useful to the geosciences and the "biosciences" in asserting their place in research on major environmental issues, which are unavoidable in the majority of calls for projects.

1.4.3. *The protection of the natural environment at the heart of all issues*

Environmental concerns are taking a prominent place, as markers of an anthropocene moment. However, although they are undoubtedly important, they are not the subject of coastal erosion and flooding risks or of natural risks in general. Natural risks are not a problem of nature, but a problem of humans confronted with natural constraints. The desire to include the protection of nature in all research topics probably responds to the strong influence of certain disciplines (such as ecology), the militant commitment of certain researchers and the political orientations of research programming. Since the 2000s, ecological concepts have been used in all subjects: the eco-socio-system of the 1990s has been transformed into a more bio-centric “socio-eco-system”, “ecosystem services” are revaluing nature economically, “nature-based solutions” are renewing the engineering sciences, and the “Anthropocene” is offering a new era for the natural sciences, which are reconnected with society.

An example of the limits of these “environmentalist” recommendations is the way in which “nature-based” solutions are presented, by the Ministry of the Environment in particular, as innovative, even revolutionary, responses to coastal risks: the idea, which is quite logical in absolute terms, is to leave buffer spaces (marshes, mangroves, coastal dunes) on the seafront to absorb the energy of storm swells and the rise in sea level. In terms of sediment dynamics, the principle is quite judicious. The problem, for the moment unavoidable, is that social dynamics have made these sectors bordering the sea very coveted spaces, often occupied by constructions of great economic value protected by hard structures that should be deconstructed to let the natural dynamics play. This is the idea of relocation which, as we have seen, comes up against strong social reticence and problems of public expenditure, which have been insurmountable until now. Thus, the heart of the problem is not to understand natural dynamics, but the social dynamics that have strongly transformed the coasts. These nature-based solutions are quite applicable, but on natural spaces, which *a priori* are not or hardly threatened by coastal risks.

1.4.4. *The use of nature as a compelling argument*

Presenting coastal risks under the prism of environmental protection leads to an amalgam that some elected officials, and especially the associations of local residents and second homeowners who are looking after their privileges, know how to use skillfully. Environmental considerations are used as irrefutable arguments. Protection of nature and protection against the sea are presented as one and the same fight, whereas they are almost opposite positions. By making nature protection a priority issue, the notion of the Anthropocene does not help to disentangle particular interests from the general interest in coastal risks.

1.4.5. *A social utility?*

Although the notion of the Anthropocene does not seem to have any particular innovative contribution, it can have real social utility as a propitious moment for paradigm inflection. In fact, it is "fashionable" and attracts media attention and, in this sense, it can encourage a more respectful occupation of the coasts of natural dynamics and raise awareness of the necessary measures of adaptation in a context of climate change. However, this potential lever for action is not without major drawbacks, as already mentioned.

1.5. References

Berque, A. (1996). *Être humain sur la terre : principes d'éthique de l'écoumène*. Gallimard, Paris.

Birkmann, J. (ed.) (2006). *Measuring Vulnerability to Natural Hazards: Towards Disaster Resilient Societies*. United Nations University Press, Tokyo.

Buchou, S. (2019). Quel littoral pour demain ? Vers un nouvel aménagement des territoires côtiers adapté au changement climatique. Report submitted to the French Prime Minister and the Minister for Ecological and Inclusive Transition, Paris.

Cabantous, A., Lespagnol, A., Péron, F. (2005). *Les Français, la terre et la mer*. Fayard, Paris.

Cazaux, E., Meur-Ferec, C., Peinturier, C. (2019). Le régime d'assurance des catastrophes naturelles à l'épreuve des risques côtiers. Aléas versus aménités, le cas particulier des territoires littoraux. *Cybergeo: European Journal of Geography* [Online]. Available at: http://journals.openedition.org/cybergeo/32249 [Accessed 26 May 2020].

Cazenave, A. and Llovel, W. (2010). Contemporary sea level rise. *Annual Review of Marine Science*, 2, 145–173.

Cerema. (2018). Cartographie nationale des ouvrages et aménagements littoraux [Online]. Available at: www.geolittoral.developpement-durable.gouv.fr/cartographie-nationale-des-ouvrages-et-r502.html [Accessed 26 May 2020].

CGEDD (2019). Recomposition spatiale des territoires littoraux. Report, Conseil général de l'environnement et du développement durable, Paris.

Corbin, A. (1988). *Le territoire du vide, l'Occident et le désir de rivage 1750–1840*. Flammarion, Paris.

Cutter, S.L., Boruff, B.J., Shirley, W.L. (2003). Indicators of social vulnerability to environmental hazards. *Social Science Quaterly*, 84, 242–261.

D'Ercole, R. (1994). Les vulnérabilités des sociétés et des espaces urbanisés : concepts, typologie, modes d'analyse. *Revue de Géographie Alpine*, 4(LXXXII), 87–96.

D'Ercole, R. and Pigeon, P. (1999). L'expertise internationale des risques dits naturels : intérêt géographique. *Annales de géographie*, 608, 339–357.

D'Ercole, R. and Pigeon, P. (2000). L'évaluation du risque à l'échelle internationale. Méthodologie et application aux diagnostics préalables aux actions de préparation et de prévention des catastrophes. *Cahiers savoisiens de géographie*, 1, 29–36.

GIEC (Groupe d'experts intergouvernemental sur l'évolution du climat) (2013). *Climate Change 2013: The Physical Science Basis. Contribution of Working Group I to the fifth Assessment Report Panel on Climate Change*. Cambridge University Press, Cambridge.

GIEC (Groupe d'experts intergouvernemental sur l'évolution du climat) (2018). Les scénarios +1,5 °C. Special report, Groupe d'experts intergouvernemental sur l'évolution du climat, Geneva.

Gilbert, C. (2009). La vulnérabilité : une notion vulnérable ? À propos des risques naturels. In *Risques et environnement : recherches interdisciplinaires sur la vulnérabilité des sociétés*, Becerra, S., Peltier, A. (eds). L'Harmattan, Paris.

Hellequin, A.-P., Flanquart, H., Meur-Ferec, C., Rulleau, B. (2013). Perceptions du risque de submersion marine par la population du littoral languedocien : contribution à l'analyse de la vulnérabilité côtière. *Nature, Science, Société*, 21, 385–399.

Hénaff, A. and Philippe, M. (eds) (2014). Gestion des Risques d'érosion et de Submersion Marines. Methodological guide, Projet-ANR-10-CEPL-0001Cocorisco.

Hénaff, A., Meur-Ferec, C., Lageat, Y. (2013). Changement climatique et dynamique géomorphologique des côtes bretonnes. Leçons pour une gestion responsable de l'imbrication des échelles spatio-temporelles. *Cybergeo: European Journal of Geography* [Online]. Available at: http://cybergeo.revues.org/26058 [Accessed 26 May 2020].

Meur-Ferec, C. and Guillou, E. (2020). Interest of social representation theory to grasp coastal vulnerability and to enhance coastal risks management. *PsyEcology*, 11(1), 78–89.

Meur-Ferec, C. and Morel, V. (2004). L'érosion sur la frange côtière : un exemple de gestion des risques. *Nature, Science et Société*, 12, 263–273.

Meur-Ferec, C. and Rabuteau, Y. (2014). Plonevez-Les-Flots : un territoire fictif pour souligner les dilemmes des élus locaux face à la gestion des risques côtiers. *L'Espace Géographique*, 43(1), 18–34.

Meur-Ferec, C., Deboudt, P., Morel, V. (2008). Coastal risks in France: An integrated method for evaluating vulnerability. *Journal of Coastal Resarch*, 24(2), 178–189.

Meur-Ferec, C., Le Berre, I., Cocquempot, L., Guillou, E., Hénaff, A., Lami, T., Le Dantec, N., Letortu, P., Philippe, M. (2020). Une méthode de suivi de la vulnérabilité systémique à l'érosion et la submersion marines. *Développement Durable et Territoires*, 11(1) [Online]. Available at: https://journals.openedition.org/developpementdurable/16731 [Accessed 16 June 2020].

Michel-Guillou, E. and Meur-Ferec, C. (2017). Living in an "at risk" environment: The example of so-called "natural risks". In *Handbook of Environmental Psychology and QOL Research*, Fleury-Bahi, G., Pol, E., Navarro, O. (eds). Springer, New York.

Mineo-Kleiner, L. and Meur-Ferec, C. (2016). Relocaliser les enjeux exposés aux risques côtiers en France : points de vue des acteurs institutionnels. *[VertigO] La revue électronique en sciences de l'environnement*, 16(2) [Online]. Available at: http://vertigo.revues.
org/17656 [Accessed 26 May 2020].

Ministère de l'Écologie, de l'Énergie, du Développement durable et de la Mer (2010). *La gestion du trait de côte*. Quæ, Paris.

Ministère de l'Écologie, du Développement durable et de l'Énergie (2012). Stratégie nationale de gestion intégrée du trait de côte : vers la relocalisation des activités et des biens. Direction générale de l'Aménagement, du Logement et de la Nature, Paris.

Ministère de la Transition écologique et solidaire (2017). Des pressions plus fortes en bord de mer, surtout dans les territoires ruraux et périurbains. Commissariat général au développement durable, Paris.

Moscovici, S. (2001). Why a theory of social representations? In *Representations of the Social: Bridging Theoretical Traditions*, Deaux, K., Philogène, G. (eds). Blackwell Publishing, Malden.

Neumann, B., Vafeidis, A.T., Zimmermann, J., Nicholls, R.J. (2015). Future coastal population growth and exposure to sea-level rise and coastal flooding: A global assessment. *PLoS ONE*, 10(3) [Online]. Available at: https://journals.plos.org/plosone/article?id=10.1371/journal.pone.0118571.

Nicholls, R.J. and Cazenave, A. (2010). Sea-level rise and its impact on coastal zones. *Science*, 328(5985), 1517–1520.

Noin, D. (1999). La population des littoraux du monde. *L'information géographique*, 2, 65–73.

November, V. (2002). *Les territoires du risque. Le risque comme objet de réflexion géographique*. Peter Lang, Berne.

ONU (2019). Sand and sustainability: Finding new solutions for environmental governance of global sand resources. *UNEP*, Geneva [Online]. Available at: https://unepgrid.ch/storage/app/media/documents/Sand_and_sustainability_UNEP_2019.pdf.

Perherin, C., Meur-Ferec, C., Deniaud, Y. (2017). Une typologie des processus de concertation lors de l'élaboration des cartes d'aléas littoraux. *[VertigO] La revue électronique en sciences de l'environnement*, 17(3) [Online]. Available at: http://journals.openedition.org/vertigo/18750 [Accessed 26 May 2020].

Pigeon, P., Rebotier, J., Guézo, B. (2018). Ce que peut apporter la résilience à la prévention des désastres : exemples en Lavours et en Chautagne (Ain, Savoie). *Annales de géographie*, 719(1), 5–28.

Rebotier, J. (2012). Une approche territoriale des risques. *Géographie et cultures*, 81, 77–90.

SOeS (Service de l'observation et des statistiques) (2013). Prix des terrains à bâtir sur le littoral métropolitain en 2011. Observatoire national de la mer et du littoral.

Toulier, B. (ed.) (2016). *Tous à la plage ! Villes balnéaires du XVIII*[e] *siècle à nos jours*. Cité de l'architecture et du patrimoine, Lienart, Paris.

Trope, Y. and Liberman, N. (2010). Construal level theory of psychological distance. *Psychological Review*, 117, 440–463.

Twigger-Ross, C.L. and Uzzell, D.L. (1996). Place and identity processes. *Journal of Environmental Psychology*, 16, 205–220.

2

Forest Fires in the Anthropocene: Issues of Scale

Christine BOUISSET
Université de Pau et des Pays de l'Adour, TREE, CNRS – E2S UPPA, Pau, France

2.1. Introduction

The objective of this chapter is to examine how the Anthropocene is renewing, or not, our view of an ancient risk, fire, a risk which is widespread and has a strong impact on landscapes. A wildland fire is a blaze that develops without control in space and time. Although in some regions of the world, such as in boreal forests, lightning is the primary cause, most fires are related to human causes (Bowman *et al.* 2017). This is particularly the case in Mediterranean regions as well as in California and Australia, which have been in the news regularly for the past few years due to the scale of fires in the outskirts of large urban areas. This anthropogenic dimension can be observed from the global to the micro-local scale. In fire-prone areas, there is a close correlation with human presence: according to figures from the Ministry of the Interior (2019) in France, 80% of fires start within 50 m of homes.

Whether they are voluntarily responsible (clearing, revenge) or involuntarily responsible (carelessness, accidents) for the outbreak of fires, societies also play an important role both upstream through changes in land use and, further downstream, on the spread of fires through the management of natural areas and extinguishing actions. While this human responsibility has been known for a long time on a regional and local scale, entry into the Anthropocene and the issue of climate change

Risks and the Anthropocene,
coordinated by Julien REBOTIER. © ISTE Ltd 2021.

raise questions about human responsibility on a global scale. In this context of reflection on the interactions between societies and environments and the great fears associated with the future of the planet and humanity, fires, which provide spectacular images, often exert a morbid fascination, symbolize the idea of catastrophe and attest to the "warming" underway, as evidenced by the frequent use of the expression "the world is burning". On the basis of a review of the literature, we will show how the arrival of questions about global change has renewed the gaze on fires on an international scale, giving them a new visibility and stimulating research on the subject. We will then examine whether these questions are concretely translated into public policies in the French national context. We will show how the topic of climate change, in particular, has refocused attention on a risk that has been somewhat forgotten in recent years because it seems to be fairly well controlled. Finally, by mobilizing a survey of semi-structured interviews with elected officials (mayors and deputies) in the south of the Landes department[1], we will show that this renewal does not necessarily translate into a paradigm shift in risk management at local level. This change of scale, by recontextualizing the issue of fires within the framework of a territorial approach, allows us to better understand the genesis and implications of risks and their management (Bouisset *et al.* 2019).

2.2. The Anthropocene or the resurgence of questions about fire and firefighting

2.2.1. *Anthropocene or pyrocene?*

2.2.1.1. *On the punctual importance of fire control…*

The emergence of the issue of global change has given rise to a clear revival of interest in fire and wildfires and to several types of reflection. The first is conceptual and questions the meaning and place of fire in the Anthropocene. The second is in the tradition of analyzing ecosystem functioning and focuses more on the interactions between global changes, in particular climate change, and fires in a systemic perspective.

1 This survey work was conducted as part of the RiTTA research project, *Risques et transformations territoriales en Aquitaine*, funded by the *Conseil régional*. The survey conducted during the fall of 2018 and winter 2019 interviewed 30 elected officials of the Pays Adour Landes Océanes in the south of the Landes department. The municipalities considered are representative of the territory, that is, urban and rural municipalities, coastal municipalities and others located further inland. They are rather small (eight have less than 1,000 inhabitants, six have more than 5,000). The interviews addressed three main themes: the living environment, climate change and natural risks and their management.

Approximately 3% of the Earth's land surface burns annually (Jia *et al.* 2019). It has long been attested that fire is a central component of ecosystem functioning with the earliest evidence dating back to the Silurian (≈420 million years ago) at a time very close to the appearance of the first plants (Scott and Glasspool 2006). Its evolution during the history of the planet is correlated to that of the oxygen level in the atmosphere (Scott and Glasspool 2006). Recent work emphasizes the importance of the systemic dimension of fire and its influence on global processes as it "links and influences regional and global biogeochemical cycles, human activity, and vegetation characteristics" (Bowman *et al.* 2009); hence, there is a need to better understand the interactions between fire and anthropogenic global environmental change.

In fact, fires cannot be considered only as an ecological phenomenon. Historical analyses of the evolution of burned areas on a global scale tend to show a close link with human activities. After a decline linked to fire control policies since the 1950s, the last decades have been marked by a phase of rapid increase (Marlon *et al.* 2016) that seems to be linked, on the one hand, to land clearing in the intertropical zone and, on the other hand, to climate change.

As several authors remind us, among living beings, the hallmark of human beings is the mastery of fire. It is a powerful transformer that has facilitated the extension of human powers (Pyne 2012; Dalby 2018) and has profoundly altered terrestrial landscapes and ecosystems (O'Connor *et al.* 2011). One thinks, of course, first and foremost of the use of fire as a means of clearing, fertilizing, or maintaining land and thus as the primary vehicle for ecosystem transformation. This is why some consider that the beginning of the Anthropocene should not be dated according to the increase of CO_2 concentrations in the atmosphere generated by the industrial revolution, but from the moment when societies began to play an important role in the shaping of ecosystems. However, these approaches to the Anthropocene through the prism of ecological transformations are criticized for trivializing the impact of societies on the environment and for ignoring the fact that what characterizes the Anthropocene is not so much their transformative capacity as such, but rather the now global and systemic nature of this capacity (O'Connor *et al.* 2011). Until recently, the transformative capacities of ecosystems through human-induced fire were documented primarily at local and regional scales.

2.2.1.2. *... to the systemic and globalized effects of combustion control*

Beyond the agricultural practices linked to fire and their effects on ecosystems, it is also underlined that the transformative capacities of human beings rely more largely on the physical process of combustion. This process has proven to be indispensable to the technological development of societies, starting with metallurgy. The American environmental historian Stephen Pyne points out that the industrial revolution was above all linked to the arrival of new fuels (coal and then oil supplanted wood) and to the mastery of new combustion techniques (steam engine, combustion engine). It is also the mastery of combustion processes that ensures the passage from manual weapons to "firearms", founding military power and what the geographer Simon Dalby calls *firepower* to designate more broadly the processes linked to modernity. The latter would mark a "pyric" transition (Pyne 2012) with the passage from open hearths to the mastery of closed hearth combustions. Both Pyne and Dalby point out that these essential elements of globalization rarely attract attention when it comes to thinking about climate change and the Anthropocene, even though combustion in all its forms is largely responsible for carbon dioxide emissions. According to Pyne, human activities have reformatted the Earth's pyrogeography. Humans bring fire to places that would not otherwise burn and exhume combustible materials (coal, gas, oil) that are naturally beyond the reach of fire, transforming the period into an age of fire , and thus into a "pyrocene" (Pyne 2019), a term that is finally more precise in his eyes than that of the Anthropocene: "Climate history has become a subset of fire history. Fire is not simply filling the void of an interglacial but asserting itself with the power of a distinctive fire age. The so-called Anthropocene, the age of humans, might as aptly be termed the Pyrocene" (Pyne 2019). It is the global character of anthropogenic changes in fire regimes and pyrogeography as well as the existence of feedbacks on a global scale that would then allow us to speak of a pyrocene, with fire being considered a determinant of climate.

The fact is that, according to modeling from satellite data, fire-related CO_2 emissions (deforestation, degradation fires – forest, savanna – peat fires) account for about 25% of global emissions (Van der Werf *et al.* 2010). According to the IPCC (Intergovernmental Panel on Climate Change) data from 2007, fire-related CO_2 emissions (fires and biomass burning) are equivalent to 50% of fossil fuel emissions. Although post-fire vegetation recovery contributes to CO_2 storage, emissions from deforestation-related fires account for about 19% of the radiative forcing since pre-industrial times (Bowman *et al.* 2009). Thus, while fires are a marker of the negative effects of climate change, they are also a driver of it. This global dimension, as well as the spectacular nature of megafires, described by the philosopher Joëlle Zask (2019) as a "total event", makes fires both a symptom and a cause of the Anthropocene. Consequently, fire risk management policies are no longer just a localized issue to protect ecosystems and local populations. The issue becomes global since fires contribute to greenhouse gases, and conversely, the biomass that escapes from them ensures CO_2 storage. This is why future assessments of anthropogenic global climate forcing should include specific analyses of the role of fire (Bowman *et al.* 2009), while work is emerging that calls for addressing the issue of fire and its management from the perspective of socio-ecological systems (Fischer *et al.* 2016) in a transdisciplinary and transcalar perspective.

But it is mainly on the first aspect, the impact of climate change on the fire regime, that the main scientific works have been focused until now.

2.2.2. *Understanding the increase in fire risk as a consequence of climate change*

2.2.2.1. *From large-scale attention to the physical conditions and impact of fire…*

Since the 1950s, interest in fire has grown under the impetus of the life and earth sciences. The main works aim to better understand and model the behavior of wildfires as well as to measure their effects on ecosystems. Research work has, therefore, focused on the micro- and mesoscales to analyze fire behavior, the flammability and combustibility[2] of vegetation, and its ability to recover at species or plant formation level.

On a regional scale, climate is the main factor explaining the sensitivity of vegetation and the distribution of fires on the planet: the climate conditions short-term meteorological parameters (precipitation, humidity, temperature, wind) and it interacts with vegetation in the long term (flammability and spatial distribution of fuel[3]). Significant variability in annual fire balances is also directly correlated to interannual variations and to the presence of droughts in particular. Historically, a close link between fire on a global scale and past climate changes over the last two millennia has been established (Marlon *et al.* 2009). Therefore, it is clear that anthropogenic climate change is having an impact on fire regimes, particularly in areas affected by simultaneous increases in temperature and decreases in precipitation.

Climate change is contributing to modification of the fire regime, with the multiplication of megafires that are characterized both by their large size (thousands

---

2 The behavior of a plant is characterized by ignition speed, which varies according to species, and combustibility, the energy released by the combustion, which varies according to species but also at the scale of plant formation, depending on its density and its continuity.

3 The structure of the vegetation plays a major role in the propagation of fire, with horizontal (cover rate) and vertical (ground to treetop) continuity favoring its progression.

or even hundreds of thousands of hectares as observed in New South Wales in Australia during the 2019–2020 season) and their intensity in terms of combustion. This is the case, for example, in fire prone temperate and semi-arid zones where destructive episodes have been observed in recent years, such as in California, Greece and Portugal in 2018 or in Australia during the 2019–2020 season with nearly 19 million hectares burned (Boer *et al.* 2020). The combined consequences of climatic changes and the passage of fires are devastating: global data suggest that "tipping points" are approaching in some of the world's most important woody biomes, including savanna forests, temperate forests and boreal forests (Adams 2013). Work in these areas, therefore, seeks to refine our understanding of the mechanisms at play from the perspective of modeling and prospectively assessing the evolution of risks and ecosystems in relation to various climate change scenarios.

These developments also raise the question of how to adjust fire control and ecosystem preservation policies. The former appear to be largely obsolete, especially if they rely solely on fire suppression (Williams *et al.* 2011), since firefighting techniques are largely impotent in the face of large fires. However, the human stakes are increasing because some of the affected regions are heavily populated: the last few decades have seen the development of interface zones between habitat and natural areas as a result of urban sprawl, agricultural decline or a combination of the two processes. For example, it is estimated that the number of homes in interface areas in the United States increased by more than 41% between 1990 and 2010 (Radeloff *et al.* 2018). So much so that in 2019, more than 1.7 million homes (with an estimated value of about $540 billion) were included in high fire risk areas (CoreLogic 2019). In 2018, the Camp Fire, the most devastating fire in California's history, resulted in the deaths of 85 people and the destruction of nearly 19,000 buildings (CoreLogic 2019). Assessing the evolution of risk under the dual influence of territorial transformations and climate change, and protecting these interface areas, have consequently become major concerns and field of research (Bento-Gonsalves and Viera 2020).

2.2.2.2. ... *to the difficult explanation of the multicausality of fires on a smaller scale*

This question of the impact of climate change on the evolution of fire regimes is a major scientific issue, complicated by the uncertainties related, on the one hand, to the respective weight of climatic factors and human actions in fire mechanisms and, on the other hand, to the possible feedbacks of a change in fire regimes on the vegetation and the climate. One of the challenges at the global scale is therefore to determine more precisely the respective contribution of temperature, precipitation and human intervention in the evolution of fires in order to refine forecasting models (Harris *et al.* 2016; Hantson *et al.* 2016; Jia *et al.* 2019). In this regard, the scientific literature produces sometimes contradictory results depending on the methodologies used (charcoal deposition analysis, remote sensing, modeling, etc.), the time steps considered and the scales of observation. For example, the analysis of historical meteorological series in the western United States shows the primacy of climatic parameters (rising temperatures and early spring) over the evolution of land cover in the annual evolution of large fires on a regional scale. The strong increase in fires observed from the 1980s onwards also concerns regions with little urbanization, where changes in land use therefore play a small role (Westerling *et al.* 2006). In contrast, the latest state of knowledge proposed by the IPCC argues conversely that while on a global scale precipitation was the main driver of the fire regime in the pre-industrial era, human factors would now be predominant (Jia *et al.* 2019). These contradictory statements are only apparently so since the scale considered is not the same and strong regional contrasts may exist. The influence of human activities can be very different between a boreal forest, a savanna or the periphery of a large agglomeration. And we know that the evolution of burned areas over the last decades is very different from one region of the world to another. A second difficulty arises from the complexity of the causal relationships and the nature of what is taken into account: are we talking about fires and their triggering factor? Are we talking about the frequency of fires? Their size? Their location? Obviously, the causal links are more or less simple to establish and more or less direct. If we consider that humans are now the main factor in climate change, it is indeed possible that societies are both directly responsible on a local and short-term scale (by setting fires) and indirectly on a global and long-term scale (by climate change and the effects induced on vegetation).

Scientific work on this issue is therefore aimed at refining understanding of the mechanisms involved from the perspective of modeling and prospective assessment of risk evolution in relation to the various climate change scenarios (for modeling issues relating to the urban climate, see Chapter 3). The current state of knowledge points to a significant increase in fire risk in some areas, particularly in southern Europe, "with dramatic increases in wildfire risk that will require increased resources and management efforts for disaster prevention and recovery" (Liu *et al.* 2010).

One of the questions raised by these scenarios is how to take them into account in risk management. This will be the subject of the following section, where, using the French example, we will examine the impact of the climate change issue on expert and non-expert views of fire.

2.3. Fire management in France: a renewed interest

2.3.1. *Integrating the climate problem into fire risk management in France: modeling and forecasting*

2.3.1.1. *Management historically focused on the hazard*

In France, fire risk management policies are based on a doctrine whose main principles and some of its major turning points can be listed. After the Second World War and the catastrophic fires of the 1940s in the Landes region of France, forest fire protection was gradually structured around, on the one hand, establishment of the departmental fire and rescue services in charge of extinguishing fires and, on the other hand, the forestry sector, which was responsible for the maintenance and management of forests. The period from the 1960s to the 1990s marked the gradual modernization of this approach, which is essentially based on a strategy for fighting declared fires: attacking incipient fires and accompanying measures to implement it (surveillance, equipment, etc.) in a very hazard-centered approach (aerial methods, "risk" indices based on climatic and ecological parameters, etc.).

In 2003, the heat wave caused large-scale fires in the Southeast, resulting in deaths and destruction of homes and drew attention to the problem of interface zones between forest and habitat. This episode revealed the very poor consideration of the social and territorial dimension of a risk that is nevertheless evolving rapidly due to peri-urbanization, urban sprawl and the growing recreational use of natural areas. This awareness led the public authorities to reactivate an old regulation, the regulation of the use of fire in and around forests and, above all, the obligation to clear the undergrowth around infrastructures and dwellings.

It also led to the acceleration of forest fire risk prevention plans, which were previously very timid, if not non-existent. These first prevention plans, which were themselves very hazard-centered, often provoked an outcry from the populations concerned and from elected officials (Bouisset 2011). As the hazard is diffuse and as vegetation covers large and homogeneous areas, even entire municipalities are classified as red zones (zones of maximum constraint). These difficulties have led to a gradual evolution of official recommendations in favor of taking greater account in the prevention plans of territorial contexts and vulnerability through the notion of "stakes" and especially of the "defensibility" of these stakes, that is, the capacity of the emergency services to defend them according to their configuration (access, presence of water points, etc.). Without at all questioning the primacy of a strategy based on extinguishing incipient fires as quickly as possible, the doctrine is therefore evolving toward a less sectoral management based on the fire fighter–forester pairing to gradually integrate urban planning and development issues. The question is to know what the arrival of concerns related to climate change produces in this context.

2.3.1.2. *From climate change in hazard evolution to the multi-causality of fires*

The first public document at national level to provide a precise diagnosis of the challenges of climate change in terms of fire is an inter-ministerial report from 2010 entitled "Climate change and the extension of areas susceptible to forest fires". This report is still a reference since it is cited by all other official documents related to the issue, whether in public communication or in operational documents at regional and local levels. This report examines all aspects of fire risk management, from the evolution of the hazard, to the issues at stake and the effectiveness of public policies. Like all expert reports on climate change, it has a strong prospective dimension as it examines the foreseeable evolution of each of these components with particular attention to climatic factors and their influence on fire behavior and vegetation sensitivity. While the report is intended to be a comprehensive overview that includes the hazard aspect as well as the socioeconomic aspects (it insists in particular on the development of wildland urban interfaces) and public policies, the most widely circulated part of the document focuses on the hazard and concerns the evolution of sensitive areas. In particular, one finds in most of the subsequent public documents a map showing the worsening of the hazard in the currently sensitive areas of southern France (Mediterranean and Landes) and the future progression of the hazard toward the north. The forecast, which is based, among other things, on the assumptions of the model from Météo-France ARPEGE-Climat, states that sensitive areas (about one-third of the forest areas in mainland France today) could increase by 30% by 2040.

By 2050, nearly half of the moors and forests in mainland France could be affected by a high level of vegetation fire hazard (Chatry *et al.* 2010). The report emphasizes the intrinsic methodological difficulty of estimating future risks, since it is based on regional-scale climate models, at best (the sensitivity of the ARPEGE-Climat model is in the order of 50 km), when the current fire hazard map, as used in the daily assessment of the level of risk or that of the prevention plan, is rather of the order of 1:25,000. This shows the difficulty of providing a fine assessment that can be directly used at the operational level. Moreover, the report emphasizes the uncertainty regarding the future of forests and natural areas as a function of bioclimatic parameters, but also of the evolution of land use (division of natural–agricultural–urbanized areas) and of the management policies applied to forests (political choices made, for example, with regard to energy transition).

In addition to the issues of knowledge and updating the forecasting model according to the progress of the IPCC's work (this is the only reference in the report to the international context – apart from European cooperation), the report recommends that climate change be taken into account immediately in various existing aspects of forest defense, for example, by developing operational documents at departmental level in areas likely to be newly confronted with this risk or by already integrating it into land-use planning "given the lifespan of certain structures (equipment, housing)". It insists on the role of prefects in knowledge transfer to the communes and reaffirms the relevance of the prevention plans for the most exposed because "they are levers to obtain from the local authorities more precautionary and if necessary corrective development policies".

Subsequent documents, such as the 2016 assessment mission on forest fire defense (Makala *et al.* 2016), take up the same elements, develop them and thus integrate without revolution the issue of climate change into the traditional aspects of fire management, that is, prevention, forecasting and control, in order to take into account the worsening of the hazard in areas already affected and its extension to territories previously spared. As on an international scale, the knowledge and recommendations that result from this are being refined, while at the same time highlighting the conceptual and methodological difficulties of integrating environment–society interactions into the understanding of risk: human factors in the triggering of fires "blur the climatic signal"[4] just as fire-fighting policies "distort" analysis of the evolution of wildfires. At the operational level, the recommendations are beginning to have an impact, particularly in terms of geographical distribution to areas that have not been affected to date. For example, in 2019, the Pyrénées-Atlantiques department has been provided with a departmental plan for the protection of forests against fire, justified by the particular problem of excessive agricultural burnings in mountainous areas and the increased risk of fires due to climate change.

2.3.2. *A recognized theme in the reflection on climate change*

2.3.2.1. *A growing institutional and media echo*

In parallel with this integration of the issue of climate change into sectoral fire management, fires are also part of the national strategy to combat climate change. For example, fires appear in the two national plans for adaptation to climate change with increasing importance: in the first national adaptation plan of 2011–2015, they

4 www.meteofrance.fr/climat-passe-et-futur/impacts-du-changement-climatique-sur-les-phenomenes-hydrometeorologiques/changement-climatique-et-feux-de-forets [Accessed 16 June 2020].

were included in the action sheet "natural risks" but in the second plan of 2018–2022, they are the first theme of intervention in the context of the action area "prevention and resilience" and are the subject of a dedicated action. Logically, this action takes up the major orientations defined in the sectoral documents.

This growing importance is related to increased media coverage of the fire-climate change link. Since the development of the press, especially regional press, fires have indeed been a recurrent subject for headlines and spectacular images. Since the beginning of the 2010s, the media have been focusing on major fires that have occurred abroad, particularly in California and Australia, and which have raised recurrent questions about climate change. Although the link between a particular fire and a particular drought episode seems easy to demonstrate, these questions are in line with broader questions about the attribution of extreme phenomena to climate change. The latter have become a veritable chestnut since extreme climatic phenomena are now almost always the subject of questions, or even assertions, about responsibility for climate change. Thus, to take a recent example, the newspaper *Le Monde* was quite affirmative in an article published on February 13, 2020 based on an AFP dispatch: "Fires strike Australia every year at the end of the southern winter. But this year they were particularly early and intense due to drought and high temperatures, phenomena fueled by climate change".

2.3.2.2. *An explicit link between sectoral and global issues*

On another scale, the case of the Pyrénées-Atlantiques is interesting because through media coverage of the issue of fine particle pollution generated by fire and its health impact, it allows us to see the two-way traffic and the porosity between sectoral concerns and transversal concerns, between local concerns and global concerns. In February 2019, an air pollution alert in the department led the prefecture to temporarily suspend agricultural burning permits for health reasons. At the same time, the episode gave rise to mobilization. For example, a resident of the Aspe valley initiated an online petition entitled "Agricultural burning in the Pyrenees, we are suffocating", which gathered more than 2,000 signatures to demand the intervention of the public authorities so that "the population be informed of the risks incurred for their health (in particular that of children and people at risk) and that this practice be better supervised". And in addition to the local issue, one of the signatories gives a global dimension to his mobilization by linking it to the international negotiations on climate: "Enough of these huge polluting and destructive fires: when we pretend to give lessons on climate to the whole world with the Paris agreement, we clean up our own backyard..." The emergence of this concern is concomitant with the deployment of devices for measuring air pollution as part of the national climate-air-energy policy, which reveals, sometimes incidentally, that fires and burning present a health risk[5]. In return, the polluting emissions of fires justify all the more the need to fight them. The greenhouse gas argument is making its way into operational documents, such as the Aquitaine interdepartmental forest fire protection plan[6] updated in 2019: "The Forest Fire Defense allows the forest to fully assume its role as a carbon pump by participating in the protection of the forest, which stores carbon, and by avoiding massive greenhouse gas emissions from fires" (PidPFCI 2019).

At the same time, the cumulative effect of the evolution of various risks affected by climate change is also becoming a topic of reflection. Still in the PidPFCI, the interactions between fire and hydro-meteorological phenomena are listed in a double entry table that evokes the way in which various fire risk parameters (ignition, flammability, combustibility, seasonal amplitude and access for rescue services) are modified by other hazards that are themselves evolving under the effect of climate change: thunderstorms (lightning strikes), storms (windthrow) and floods (accessibility problems). This makes management all the more complex and raises questions about how the territorial dimension of risk is taken into account and how strategies promoted at higher levels are implemented in the territories.

2.4. Fires, climate change and territory: a mobilizing subject?

2.4.1. *The perverse effects of over-efficient crisis management*

2.4.1.1. *A strategy focused on the treatment of the hazard rather than the causes*

In 2016, the interministerial evaluation mission recommended structuring the national strategy around four main principles, the first being a "principle of awareness" of the risk and the seriousness of the threat (Makala *et al.* 2016). While noting the reduction in burned areas, she judges that "the probability of a large-scale fire, with dramatic consequences, remains significant. The number of fires ignited has decreased little and the structural causes of fires remain for a large part" (Makala *et al.* 2016). The mission therefore underlines the fundamental flaw in the control

5 For example: "Since the beginning of 2015, one of the measurement stations of Marseille AtmoSud has been equipped with a battery of equipment that allows it to distinguish, among the particles, what is related to car traffic, i.e. fossil fuels, and what is related to smoke from plants. The progress of wood energy in Provence, and the more important recourse of households to wood heating, led the Association Agréée pour la Surveillance de la Qualité de l'Air to make this choice. Today it allows us to be aware of the pollution generated by forest fires" www.atmosud.org/actualite/quand-le-feu-de-foret-pollue.

6 The plan concerns the departments of the former Aquitaine region.

strategy adopted in France: the weakness of prevention in the strict sense, that is, a policy aimed at reducing the number of fires ignited by acting on the causes, in favor of a policy of extinction that seeks to reduce the surface area burned by fires outbreaks using effective but costly control means. Thus, the extinction policy represents about two-thirds of the budget devoted to fire management, 60% of which is financed by local authorities (Chatry *et al.* 2010). Prevention therefore covers about one-third of the expenditure. In reality, however, the term prevention is understood in a very broad sense: a large part of this budget is not aimed at suppressing fire outbreaks but at equipping forest areas to facilitate the intervention of emergency services (trails, water points, etc.).

In addition to its cost, the first disadvantage of this policy is that although it usually succeeds in limiting the spread of fires, it does not prevent a large fire from getting out of control because of particularly complicated natural conditions (extreme drought, violent winds, access difficulties), because the available resources have been exceeded by other demands, or even because of errors in crisis management. However, if a small fire can be quickly controlled, large fires are very difficult, if not impossible to control.

While fire outbreaks remain numerous, or even increase, the reduction in burned areas makes the risk invisible both to the general public and to certain local elected officials, who tend to minimize the risk precisely because of the absence of major fires (Bouisset and Bordier 2019). Some, nevertheless, emphasize the cost of this policy, particularly the financial contribution of the communes to the operation of the departmental fire and rescue service. This view of risk reflects territorial and demographic changes: it is often in coastal municipalities or those that have become peri-urban that municipal teams, from socio-professional categories less directly involved with forestry issues, consider that, apart from their financial contribution, risk is not really their responsibility, even though their territory is largely forested. However, measures aimed at raising public awareness and taking into account the risk of fire in land-use policies, and in particular in urban planning documents, have slowly developed since 2003 and its scorching summer.

2.4.1.2. *The problematic decoupling of territorial dynamics and fire risk*

In addition to urban planning documents, by virtue of their police powers, mayors are required to enforce mandatory brushwood clearing around dwellings and to take the place of defaulting owners by having the work done automatically. These measures are regularly indicated as priorities by specialists and technical reports. But in the Landes region, progress is modest and the means are always allocated to the fight. Above all, as the effectiveness of the latter makes the risk less worrying, some communes rely totally on foresters and firemen to avoid disaster. This invisibility obtained through forest surveillance and massive emergency response also backfires on the strategy itself: the absence of spectacular fires for a few years makes the subject less of a priority in the eyes of decision-makers and is reflected in a cyclical erosion of the resources allocated, until a large fire once again focuses attention on the subject for a while (Bouisset 1998). The 2016 evaluation mission regretted the decrease in funding and insisted on the urgency of investing in the renewal of hoses.

The other flaw in this strategy is therefore that it makes fire risk solely a question of crisis management and a matter for professionals (Chatry *et al.* 2010), foresters and firefighters, by decoupling, in the eyes of non-specialists, the question of risk from that of land use and daily life. While many of the Landes elected officials questioned about the evolution of their territory mention population growth and the strong progression of urbanization (only four elected officials do not mention them), and even though all of them know that many fires are of anthropogenic origin, none of them addresses the question of the possible links between these phenomena (increased frequentation, fragmentation of forest property, multiplication of habitat–forest interfaces) and the evolution of risk. As we have shown in another publication (Bouisset and Vanneufville 2018), this makes it all the more difficult to integrate it into urban planning policies and to accept constraints related to the protection of urbanized areas such as the obligation to clear brush around dwellings.

2.4.1.3. *On the ground, evanescent links with climate change*

In this context, where the reality of the current risk already seems remote, the question of the effects of climate change and the future evolution of fire risk seems even more difficult to grasp. The results of the survey of mayors show contrasting levels of awareness, with the number of elected officials considering the idea of a worsening of the hazard equaling the number of those who did not consider, or at least not spontaneously, the effect of climate change on fires. As with other risks, questions about the causal links and the imputability of the phenomena are also formulated by questioning the respective parts of climate change, normal climate variations and the maintenance of natural areas. The highly publicized fires in California and Australia are often taken as references, both as emblematic examples of the effects of climate change and as elements of comparison to relativize the seriousness of the local risk.

Difficult to measure, the irruption of the issue of climate change may also have another perverse effect, that of further minimizing the already weak understanding of the social component of risk at the local level by reinforcing the idea that the climate is the only important parameter: if the hazard becomes more severe, it would be inevitable that the risk would worsen in proportion. More than ever, salvation would therefore lie in the protection of people and property (urbanized) against declared fires. While national directives and the discourse of experts on risk seek both to better take into account the issue of fire in land use planning and to raise awareness of the effect of climate change on the hazard, in the field the messages are not sufficiently convincing, outside the sphere of forestry, as long as experience of fire is lacking. Especially since fire is not, far from it, the only subject of concern.

2.4.2. *The competition of multiple risks*

2.4.2.1. *The risk of fire overshadowed*

In Aquitaine, the storms of 1999 and especially that of 2009, which ravaged the region, triggered questions about the link between climate change and extreme phenomena, giving rise to a great deal of reflection and a multiplicity of initiatives, expert reports, forecasts, etc. Among these is a 2012 report by the Regional Economic, Social and Environmental Council entitled "L'Aquitaine face aux risques naturels: mieux connaître et prévenir, pour ne pas subir" (Aquitaine facing natural risks: better knowledge and prevention to prevent suffering), which highlights the diversity and multiplicity of risks present in the region. A lexical analysis of the report shows that flooding (the term actually encompasses floods and submersion problems) is the most frequently cited risk (with 217 occurrences), followed by fire (111), storms (83), land movements (81) and erosion (66). And when elected officials are asked generally about the risks affecting their commune, flooding is also widely cited: it is mentioned 19 times (according to the departmental file on major risks, only 16 of the 30 communes studied are exposed), whereas forest fires, often forgotten, are cited spontaneously by only eight elected officials, but in fact concern 18 communes. When asked whether they consider their commune to be vulnerable to climate change, half of the elected officials responded positively, again spontaneously citing the problem of flooding (10 mentions) much more than fires (three mentions).

The early 2010s also saw the formation of the regional IPCC that is AcclimaTerra, which, at the request of the regional council, produced two expert reports in 2013 and 2018 on the impact of climate change in the region. These documents emphasized the link between climate change and hazards. In addition, the damage caused by storm Klaus (2009) and the equally visible nature of erosion on certain parts of the Atlantic coastline have focused attention due to their spectacular nature and immediate economic impact. One thinks, for example, of the media coverage of the situation of the building Le Signal in Soulac threatened by marine erosion which, in the regional daily newspaper *Sud Ouest* alone, has been the subject of no less than 68 articles since 2010.

As a result, in coastal communities, the issue of climate change is primarily addressed by elected officials through the risks of flooding and coastal erosion issues. Some also point to a change in hydrological risks, which are considered more frequent and more intense. It is not surprising, therefore, to note that it is these new hazards (or at least those perceived as such), which are more visible and perhaps less known and controlled, that seem to be the most worrying.

2.4.2.2. *Climate change and future risks: a selective recognition*

Emerging natural or health risks (such as the appearance of the tiger mosquito) are clearly linked to climate change, as are changes in certain existing risks. Nevertheless, while some consider these phenomena to be visible and spectacular, and even if the elected officials do not question the existence of climate change, a large majority testify to the difficulty of understanding it because of its complexity and the abundance of information that is often contradictory. Nearly half of them question the causal links and the perceptibility of changes, either by evoking the cyclical nature of hazards, or by questioning the experience that can be made on an individual scale, around three arguments. The first is the questioning of memory (the frequent "we have never seen that"). The second is the causal links between climate change and a given event. Finally, the third argument is that of the scale of observation: while some elected officials believe they perceive the evolution of hazards, others, without denying the existence of climate change in general, doubt that what happens on a global scale and over a long period of time can be observed on a local scale and in the short term.

Consequently, while fire management policy at national and regional levels is largely considered from a sectoral perspective, local elected officials think of the future primarily in terms of territorial issues and multiple risks. Fire is not isolated from the other types of risk that affect the territory. In municipalities where several hazards are present, this leads elected officials to make comparisons, generally to the detriment of the fire risk. For unlike fire and its potential aggravation, the elected officials say they have direct experience of phenomena linked to rising water levels. These phenomena give credit to the impact of climate change in their eyes. Consequently, it is not surprising to find that, in view of the immediate human and economic stakes, the problems of flooding and coastal erosion are a much greater priority than the risk of fire, which is considered to be under control and their predicted worsening appears to be more virtual.

2.5. Conclusion

In the context of an Anthropocene that "belies this great temporal, ontological, epistemological and institutional division between nature and society" (Bonneuil and Fressoz 2013), it was interesting to look at fires, for which, more than for other so-called "natural" risks, it is easy to show that they are the product of close interactions between nature and society. The question posed was that of the possible reconfiguration of the view of risk in the context of the Anthropocene. At international level, the success of the concept has fueled a renewed interest in the theme with the development of work on the interactions and feedback loops between fire, climate and, to a lesser extent, societies. As this type of approach is already established, it is mainly through a change of scale (interactions on a global scale) and temporality (retrospective approaches are now joined by prospective work on future hazards) that this work differs from that carried out before the invention of the Anthropocene. Apart from a few works by historians, anthropologists or geographers, the role of societies is often questioned more by the way they manage fires from an operational perspective (how to extinguish the fire, how to protect oneself) than by the reason for the fires (their territorial, socio-political and economic context) from local to global.

In France, this proliferation of ideas has led to the progressive integration of the issue of climate change into the national strategy for the protection of forests against forest fire based on the announcement of a temporal and spatial worsening of the hazard. By broadening the functions attributed to ecosystems (one thinks of the success of the notion of ecosystem service and the abundant discourse developed on the role of CO_2 storage), contemporary environmental concerns enrich the discourse of justification of fire management policies. However, the importance of interactions is emphasized over and over again, the perspective still remains mostly sectoral and hazard centric. These interactions increase scientific uncertainty and, in the eyes of the experts, make the task of evaluation more difficult. But in themselves, they do not give rise to reflection or debate. To use Amy Dahan's (2013) expression, climate change as incorporated into fire management in France is indeed an object of science, an object of expertise, but not a political problem. As with other hazards (see Chapter 1), while it has the merit of bringing fires back to the forefront, one of the disadvantages of using the global change framing is to refocus attention on the hazard and thereby, by returning the risk to a global scale over which the local has no control, to further renaturalize an already weakly politicized problem (Reghezza 2015) even though 90% of the fires observed in France are anthropogenic. While the local level is being given increasing weight in the fight against climate change, the fire risk management strategy still faces two pitfalls that feed the "dragons of inaction" (Gifford 2011): on the one hand, cognitive limitations and optimism bias, and on the other hand, lack of consideration for the divergence of interests and trade-offs that cross territories and guide individual and collective choices at the local level. This explains why, despite scientific warnings and strategic directives from higher levels, the subject remains poorly mobilized on the ground, especially when, as in the Landes, it is competing with other risks, more immediate, more visible and less commonplace.

2.6. References

Adams, M.A. (2013). Mega-fires, tipping points and ecosystem services: Managing forests and woodlands in an uncertain future. *Forest Ecology and Management*, 294, 250–261.

Bento-Gonçalves, A. and Vieira, A. (2020). Wildfires in the wildland-urban interface: Key concepts and evaluation methodologies. *Science of the Total Environment*, 707, 135592.

Boer, M., Dios, V., Bradstock, R. (2020). Unprecedented burn area of Australian mega forest fires. *Nature Climate Change*, 10, 1–2.

Bonneuil, C. and Fressoz, J.-B. (2013). *L'événement anthropocène : la Terre, l'histoire et nous*. Le Seuil, Paris.

Bouisset, C. (1998). Incendies de forêts méditerranéens : limites, marges, frontières. PhD Thesis. Université d'Avignon et des Pays de Vaucluse, Avignon.

Bouisset, C. (2011). PPR, urbanisation et risques d'incendie de forêt dans les Pyrénées-Orientales : méthodes, enjeux, débats. *Cybergeo: European Journal of Geography*, 551 [Online]. Available at: https://journals.openedition.org/cybergeo/24658 [Accessed 12 October 2020].

Bouisset, C. and Bordier, S. (2019). Regards d'élus sur les évolutions territoriales, les risques naturels et les changements climatiques dans le Pays Adour Landes Océanes. Research report, Projet Risques et Transformations Territoriales en Aquitaine (RiTTA), Aquitaine.

Bouisset, C. and Vanneufville, S. (2018). Pourquoi être résilient quand on est efficace ? La gestion du risque incendie de forêt dans les Landes de Gascogne au défi des changements territoriaux. *[VertigO] La revue électronique en sciences de l'environnement*, Special edition 30 [Online]. Available at: https://doi.org/10.4000/vertigo.19152 [Accessed 9 May 2020].

Bouisset, C., Clarimont, S., Rebotier, J. (2019). L'environnement : une catégorie au faible potentiel mobilisateur dans un territoire industriel à risques – le cas du bassin de Lacq. In *Penser le gouvernement des ressources naturelles*, Busca, D., Lewis, N. (eds). Presses Universitaires de Laval, Quebec.

Bowman, D., Balch, J., Artaxo, P., Bond, W.J., Carlson, J.M., Cochrane, M.A., D'Antonio, C.M., Defries, R.S., Doyle, J.C., Harrison, S.P. *et al.* (2009). Fire in the earth system. *Science*, 324(5926), 481–484 [Online]. Available at: https://doi.org/10.1126/science.1163886 [Accessed 16 June 2020].

Bowman, D., Williamson, G., Abatzoglou, J., Kolden, C., Cochrane, M., Smith, A. (2017). Human exposure and sensitivity to globally extreme wildfire events. *Nature Ecology & Evolution*, 1, 0058 [Online]. Available at: https://doi.org/10.1038/s41559-016-0058 [Accessed 9 May 2020].

Chatry, C., Le Gallou, J.Y., Le Quentrec, M., Lafitte, J.J., Laurens, D., Creuchet, B., Grelu, J. (2010). Changement climatique et extension des zones sensibles aux feux de forêts. Interministerial mission report [Online]. Available at: https://www.vie-publique.fr/sites/default/files/rapport/pdf/104000494.pdf/.

CoreLogic (2019). The 2019 wildfire risk report. Report, CoreLogic, Irvine, CA.

Dahan, A. (2013). Le changement climatique : l'exception d'un risque. In *Du risque à la menace. Penser la catastrophe*, Bourg, D. (ed.). PUF, Paris.

Dalby, S. (2018). Firepower: Geopolitical cultures in the Anthropocene. *Geopolitics*, 23(3), 718–742.

Fischer, A.P., Spies, T.A., Steelman, T.A., Moseley, C., Johnson, B.R., Bailey, J.D., Ager, A.A., Bourgeron, P., Charnley, S., Collins, B.M. (2016). Wildfire risk as a socioecological pathology. *Frontiers in Ecology and the Environment*, 14(5), 276–284.

Gifford, R. (2011). The dragons of inaction: Psychological barriers that limit climate change mitigation and adaptation. *Am. Psychol.*, 66(4), 290–302.

Hantson, S., Arneth, A., Harrison, S.P., Kelley, D.I., Prentice, I.C., Rabin, S.S., Archibald, S., Mouillot, F., Arnold, S.R., Artaxo, P. (2016). The status and challenge of global fire modelling. *Biogeosciences*, 13(11), 3359–3375.

Harris, R.M., Remenyi, T.A., Williamson, G.J., Bindoff, N.L., Bowman, D.M. (2016). Climate–vegetation–fire interactions and feedbacks: Trivial detail or major barrier to projecting the future of the Earth system? *Wiley Interdiscip. Rev. Clim. Change*, 7(6), 910–931.

Jia, G., Shevliakova, E., Artaxo, P., De Noblet-Ducoudré, N., Houghton, R., House, J., Kitajima, K., Lennard, C., Popp, A., Sirin, A., Sukumar, R., Verchot, L. (2019). Land–climate interactions. In Climate change and land: An IPCC special report on climate change, desertification, land degradation, sustainable land management, food security, and greenhouse gas fluxes in terrestrial ecosystems, Shukla, P.R., Skea, J., Calvo Buendia, E. *et al.* (eds). Report, IPCC, Geneva.

Liu, Y., Stanturf, J., Goodrick, S. (2010). Trends in global wildfire potential in a changing climate. *For. Ecol. Manag.*, 259, 685–697.

Makala, A., Gohier del Re, A., Dereix, C., Brandeis, A., Grandjean, M., Granger, Y. (2016). Mission d'évaluation relative à la défense de la forêt contre l'incendie. Ministère de l'Agriculture, de l'Agroalimentaire et de la Forêt/IGA/CGEDD/CGAAER.

Marlon, J.R., Bartlein, P., Carcaillet, C., Gavin, D., Harrison, S., Higuera, P., Joos, F., Power, M., Prentice, I. (2009). Climate and human influences on global biomass burning over the past two millennia. *Nat. Geosci.*, 2, 307–307.

Marlon, J.R., Kelly, R., Daniau, A-L., Vannière, B., Power, M.J., Bartlein, P., Higuera, P., Blarquez, O., Brewer, S., Brücher, T. *et al.* (2016). Reconstructions of biomass burning from sediment-charcoal records to improve data-model comparisons. *Biogeosciences*, 13(11), 3225–3244.

O'Connor, C., Garfin, G., Falk, D., Swetnam, T. (2011). Human pyrogeography: A new synergy of fire, climate and people is reshaping ecosystems across the globe. *Geogr. Compass.*, 5/6, 329–350.

Pyne, S.J. (2012). *Fire: Nature and Culture.* Reaktion Books, London.

Pyne, S.J. (2019). Welcome to the Pyrocene: A fire creature remakes a fire planet. *Nat. Hist.*, 127(9), 3–5.

Radeloff, V.C., Helmers, D.P., Kramer, H.A., Mockrin, M.H., Alexandre, P.M., Bar-Massada, A., Butsic, V., Hawbaker, T.J., Martinuzzi, S., Syphard, A.D., Stewart, S.I. (2018). Rapid growth of the US wildland-urban interface raises wildfire risk. *PNAS,* 115(13), 3314–3319.

Reghezza, M. (2015). De l'avènement du Monde à celui de la planète : le basculement de la société du risque à la société de l'incertitude. HDR Thesis, Université Paris 1-Panthéon Sorbonne, Paris.

Scott, A.C. and Glasspool, I.J. (2006). The diversification of Paleozoic fire systems and fluctuations in atmospheric oxygen concentration. *Proc. Natl. Acad. Sci.*, 103(29), 10861–10865.

Van der Werf, G.R., Randerson, J.T., Giglio, L., Collatz, G.J., Mu, M., Kasibhatla, P.S., Morton, D.C., De Fries, R.S., Jin, Y., Van Leeuwen, T.T. (2010). Global fire emissions and the contribution of deforestation, savanna, forest, agricultural, and peat fires (1997–2009). *Atmospheric Chem. Phys.*, 10(23), 11707–11735.

Westerling, A.L., Hidalgo, H.G., Cayan, D.R., Swetnam, T.W. (2006). Warming and earlier spring increase western U.S. forest wildfire activity. *Science*, 313(5789), 940–943.

Williams, J., Albright, D., Hoffmann, A.A., Eritsov, A., Moore, P.F., Mendes De Morais, J.C., Leonard, M., San Miguel-Ayanz, J., Xanthopoulos, G., Van Lierop, P. (2011). Findings and implications from a coarse-scale global assessment of recent selected mega-fires. *5th International Wildland Fire Conference*, Sun City, South Africa, 27–40.

Zask, J. (2019). *Quand la forêt brûle. Penser la nouvelle catastrophe écologique.* Premier Parallèle, Paris.

3

Urban Climate: Agenda and Perspectives of a Climate Risk

Julia HIDALGO
LISST, CNRS, Toulouse, France

3.1. An internationalized and interdisciplinary research topic

3.1.1. *How do we define urban climate?*

Urbanized areas disrupt atmospheric characteristics at local and regional scales by altering the exchange of heat, moisture, mass and momentum between the Earth's surface and the upper air layers. These effects are present at all spatial[1] and temporal scales. They depend on the physical structure of the city and on anthropogenic activities, especially those that generate heat. These particular climatic characteristics of urban areas constitute the so-called urban climate or microclimate.

The most widely studied and publicized phenomenon is the Urban Heat Island (UHI) (Oke *et al.* 2017). It describes differences in surface and/or air temperatures compared to a "natural" environment. Other phenomena include changes in turbulence and general airflow dynamics as the air adapts to the complexity of the urban surface, influencing the dispersion of air pollutants, cloud formation in general and the water cycle in particular.

1 From street level, for example, for turbulence, to urban or even regional level for certain phenomena such as the transport of atmospheric contaminants or heat.

Risks and the Anthropocene,
coordinated by Julien REBOTIER. © ISTE Ltd 2021.

3.1.2. *Composite risks defined by climate change and urban climate*

With regard to the climate change underway on a global scale, cities have a dual role. They play an active role in aggravating climate change because by hosting more than half of the world's population and polarizing economic activity, they concentrate high levels of greenhouse gas (GHG) emissions. But urbanized areas also have a passive role. They are subject to direct changes in the regional climate that occur on different spatial and temporal scales. This is the case for the general rise in temperatures and sea levels, the modification of rainfall patterns or the intensification of extreme events (heat waves, storms and cyclones).

Such a situation calls for policies to adapt to the expected impacts of climate change, but also policies to mitigate the impact of urban areas on the atmosphere. These policies cut across the fields of operational urban planning and risk management. They act as much on the physical materiality (artificialization of the soil, three-dimensional structure of built spaces) as on the reduction of the vulnerability of populations and economic activities by means of early warning systems or information campaigns (Table 3.1). Cities can, thus, also play an important mitigation role and are the grounds for applying many policies related to climate management at local and global scales.

Objectives	Levers for action
Increase natural cover and reduce impervious surfaces	Establishment of water surfaces, permeable soils and vegetation Control of urban expansion
Reduce the impact of sunlight in indoor and outdoor spaces	Shade systems (privacy screens, street trees) Reflective building materials Thermal insulation of buildings
Optimize wind ventilation at different spatial scales	City-wide natural ventilation corridors Orientation of streets and buildings at neighborhood level Building ventilation systems (windows)
Reduce anthropogenic heat flow	Reduction of heat emissions from industrial activities, air conditioning systems, road traffic
Reduce the vulnerability of populations, anticipating risk situations	Alert plans (Heat wave plan) Information and awareness campaigns Individual behaviors

Table 3.1. *Example of policy levers to address the risk of extreme temperatures in urban areas*

There is now a plethora of scientific literature on the adaptive capacity of cities. A review of scientific publications on measures for adapting to the UHI and to heat waves provides a solid overview of 30 years of research. Energy consumption, air pollution, outdoor ambient temperature and citizen health during summer have been the focus of most efforts, regardless of continent (Akbari and Kolokotsa 2016). The second major research focus has been the development and evaluation of materials (roofs, walls and sidewalks) to counteract the effects of UHIs (Pisello 2017). Multi-level studies, comparing the effect of multiple mitigation measures to the UHI phenomenon, are less common than studies dedicated to a single mitigation measure. This can be a hindrance to decision making as the difficulty of prioritizing actions remains (Aleksandrowicz *et al.* 2017). Recommended thermal adaptation measures for outdoor spaces should not be reduced to environmental or technological interventions on the building. Behavioral adjustments and psychological adaptation must jointly be considered (Shooshtarian *et al.* 2018). Finally, Hintz *et al.* (2018) explore the transferability of solutions proposed in the literature to address urban heat waves. Most of the case studies discussed are from North America, Europe and Australia. The solutions analyzed focus mainly on the use of green (vegetation) and blue (water) or gray (built) infrastructure. The authors consider the city administration as a pivotal element for the implementation of solutions. They also emphasize the importance of residents and local authorities as key actors in urban heat wave adaptation.

3.1.3. *Internationalization and interdisciplinarity, two hallmarks of urban climatology today*

This targeted bibliographical review suggests the extent to which urban climatology has become both a deeply internationalized and interdisciplinary field of research.

Although the dissemination of knowledge within the research community takes time, a certain degree of cohesion and maturity has been achieved. Historically, the circulation of knowledge has been based on meetings organized under the auspices of large international networks such as the World Meteorological Organization (WMO), the International Council for Building (ICB), the International Federation of Housing and Planning (IFHP), the International Society of Biometeorology (ISB), the World Health Organization (WHO) and, more recently, the International Association for Urban Climate (IAUC).

Hebbert and Mackillop (2013) have faithfully reconstructed the dissemination of this knowledge, which began in the late 1950s in Davos with an international working group on urban climate studies involving the IFHP, ISB and ICB. This group continues to exist, but it is the IAUC that has taken the lead in organizing a tri-annual conference cycle and structuring research networks on an international scale. The success of this international research collaboration, knowledge exchange and networks is explained by the fact that key figures in urban climate research have participated in these international structures in various capacities. Even today, some IAUC members are heavily involved in WMO or IBS.

While it is common to refer to the work of L. Howard on London in the early 19th century and later to the Canadian T.R. Oke as a pioneer in the field from the 1970s onwards, research describing atmospheric conditions in urban areas accelerated in the 2010s, taking advantage of the development of new numerical modeling technologies and attracting a growing number of scientists from different disciplines around the world.

Researchers in physical geography and atmospheric physics had at first largely invested in urban climatology by directing research toward the urban exterior environment, whereas architects generally concentrated more on the study of microclimatic conditions inside buildings. This work concerned both the impact of urban development on the atmosphere and the response of populations to the urban climate produced. The spatial scales covered ranged from the building or the street to the city.

A gradual thematic expansion has taken place in recent years, enriching the research field by including specialists in urban air pollution, urban hydrology, building-scale microclimatology, urban vegetation, and urban planning, among others (Oke *et al.* 2017). The notion of the Anthropocene is practically non-existent in this literature and is presented as the context or the "background" of works that are often highly technical in nature (e.g., Ching 2018). More broadly, however, there are occasional reflections among urban studies that mobilize the Anthropocene moment and atmospheric phenomena through urban ambiences and their sensitive register (Thibaud 2018).

3.2. Making urban climatology knowledge operational: a challenge for action

3.2.1. *Linking knowledge production and operational needs*

3.2.1.1. *A long-standing but tenuous and growing interest*

Urban climate researchers have made significant progress in understanding the links between the extreme spatial heterogeneity of the urban surface and changes in the overlying atmosphere called the *urban boundary layer*. Significant gaps remain in our understanding of the physical processes, but the community generally recognizes that the major challenge of the moment is to articulate the knowledge gained and the decision making in urban planning.

Efforts in this direction began very early, as evidenced by Kassner's book, *The Meteorological Basics of City Planning*, which was published in German in 1910. Hebbert and MacKillop further include a comprehensive review of urban climatology studies applied to the field of urban planning in their 2013 publication. But there are more recent initiatives within the international research community.

The 2012 International Conference on Urban Climatology (ICUC8, Dublin, Ireland) included a plenary session on applied studies. In 2015, at ICUC9 (Toulouse, France) it was clearly decided to focus "on recent scientific activities on climate change mitigation and adaptation in urban areas, as well as on transfer to urban planning and development actors". That said, the success of sessions such as "transferring urban climate knowledge to planners" or "urban design with climate" still went hand in hand with a vertical conception of information transfer. The 2018 ICUC10 in New York City, meanwhile, consolidated the momentum of applied studies with numerous sessions on urban planning and governance, comfort and impact studies. We also note that a greater number of works involve field actors, or are co-constructed with them, attesting to a certain tendency to value action research.

3.2.1.2. *Some obstacles to this articulation*

There is a great deal of applied research on urban climate. It focuses on mitigating the urban heat island, managing wind, rain and snow, and using green spaces to improve air quality and human comfort. Nevertheless, the application of scientific knowledge to urban design has not been without obstacles. The exceptions are the German-speaking countries where several experiments exist (e.g. Stock and Beckröge 1985; Horsfall 1904; Kratzer 1937; Stock 1992; Baumüller 2005, to name a few), and to a lesser extent Israel (Goldreich 2003) or Japan (Kusaka 2008; Tanaka *et al.* 2009). In retrospect, Hebbert and Mackillop (2013) highlighted some possible barriers to integration between climate knowledge and operational scale. They divide them into two categories. The first refers to trends in urban planning practices and the second to the modes of producing atmospheric knowledge.

Concerning urban planning, two key aspects are highlighted by the authors:

– modernism was the dominant movement in urban production in the 20th century. The styles of architecture and urban design associated with it put more emphasis on the scale of the building. Exterior spaces were considered only from a landscape point of view, neglecting climatic analysis;

– between the 1950s and 1970s, environmental determinism gave way to environmental constructivism, leading to a decline in interest in the impact of design on the physical environment.

Concerning the production of atmospheric knowledge, it was the disconnection between national meteorological services – which produce atmospheric data and analyses – and the land use planning professions that was emphasized:

– atmospheric measurements require observation networks that are expensive to build and maintain. These networks are dependent on national meteorological services, the only structures with adequate economic and technical means. However, the main national meteorological services are less oriented toward the urban environment than toward aviation, agriculture, weather forecasting and national defense. For many, meteorological data have historically been collected outside urbanized areas. In urban areas, the complexity of the atmosphere is such that site-specific observations and high-density measurement networks are required;

– the development of numerical modeling over the last few decades has reinforced the centralized nature of the production of atmospheric knowledge by national and international meteorological services.

To all these obstacles to the circulation and application of information, we can add the specificity of research results in urban climatology, specific to each city of study, which makes it difficult to increase generality. The latter is less a question of actions, since these emerge from very different contexts, the needs of the field and the identity of decision-makers, than of methods and tools.

Research is thus striving to transcend case studies and develop methods for standardized atmospheric or urban data production, representation and analysis. Considerable efforts have been made, for example, on protocols for studying the impact of cities on atmospheric thermodynamics and chemical composition (WMO 2017). A rich literature also exists around atmospheric measurements in urban areas and modeling at various scales performed on the basis of numerical, experimental and theoretical modeling (Dabberdt *et al.* 2004; Kanda 2006; Masson 2006; Barlow 2014).

The integration of climate knowledge and territorial planning knowledge for operational purposes presents two major challenges: to compare the climate studies and analyses produced and to extend to different contexts the tools and methods developed until now mainly in the North. The use (and testing) of observation, modeling and formalization tools – often in cartographic form – is a major challenge in the context of the strong urbanization of developing countries.

3.2.1.3. *Putting climate change on the agenda, a window of opportunity to remove certain obstacles*

The laborious and costly nature of local-scale climate studies is now mitigated by the progressive development of lightweight sensors or open urban databases that allow for high-resolution mapping, and by geographic information systems (GIS) platforms that can be linked to multiscale statistical simulation or physical modeling models. Collaborative data are also strongly penetrating urban climate research, counteracting the lack of atmospheric measurements discussed above (Chapman 2015).

But the inclusion of applied urban climatology in the urban design and planning agenda must above all be linked to the growing awareness of anthropogenic climate change in public and political opinion (Bulkeley and Betsill 2003), particularly in the last decade.

The two "Cities and Climate Change" reports published in 2011 by the United Nations and the Urban Climate Change Research Network (UN 2011; UCCRN 2011) highlight the role that cities should play in the coming years to mitigate and adapt to climate change. These initiatives have played a major role in placing urban-scale climate issues on the local, national and international agenda. This is the time of the emergence of networks of cities committed to fighting climate change at international and national levels. Networks such as the C40, founded in 2010 and which now includes about 100 cities, ICLEI (Local Governments for Sustainability) and the Global Cool Cities Alliance have emerged.

The evolution of the climate on a planetary scale is thus becoming a window of opportunity for transferring climate knowledge to the local level, in particular through a strong evolution of the regulatory framework. Mitigation and adaptation are becoming explicit objectives of local urban plans in many Western countries. In addition, there is an increasing number of dedicated tools such as "Climate Plans" that establish a diagnosis and actions through mitigation and adaptation to climate change (Reckien *et al.* 2019; Grafakos *et al.* 2020). In addition to these specific tools, there is a myriad of environmental charters and plans that serve as a reference even if they are only incentive based (Lambert-Habib *et al.* 2013).

3.2.2. *Tools to facilitate management and applications*

3.2.2.1. *Outline of an applied research horizon*

Three research areas are currently focusing on the efforts of urban climatologists (Hidalgo *et al.* 2019): urban data production, analysis methods and impact indicators, and methods and tools to facilitate the articulation between climate and urbanistic knowledge.

An adequate description of cities is necessary at all stages of the process of studying urban climate and evaluating adaptation strategies to climate change. This description of the urban surface oriented toward climate studies encounters problems related to the spatial scale and the nature of the data. The emergence of a typological classification of the urban surface allows a certain standardization of this description. This typology defines Local Climate Zones (Stewart and Oke 2012) according to land use, building density, height and function. An international initiative led by the IAUC called WUDAPT (World Urban Database and Access Portal Tools)[2] aims to produce maps of most major cities in the world according to this standard. This conceptual approach, in addition to being used for micro-climatic analysis and numerical simulations, has proven to be an effective way to bring climate scientists and city practitioners together, as the typological approach is commonly used in urban planning and architectural studies.

The methods of analysis and production of impact indicators for planning and evaluating both adaptation and mitigation strategies still lack unity. The difficulty stems from the complex nature of urban functioning, and therefore from the different aspects that must be taken into account simultaneously to provide systemic and relevant information for the field of urban planning. This requires interdisciplinary and transversal analyses that always come up against micro-climatic, regulatory and local governance specificities, making the construction of common practices even more complex. Researchers are striving to define multi-criteria indicators (related to urban climate, energy consumption, population well-being, water resources) that can then be evaluated, for example, through observations and measurements or modeling that allow for exploration of the present moment or for prospective studies. These quantitative approaches lack for the

2 www.wudapt.org/ [Accessed June 15, 2020].

moment in our community a certain critical distance as to their capacity to provide a systemic vision of the issues. For example, how can we measure intangible aspects such as comfort and well-being, for which climate aspects are just one more input among others?

Finally, a careful study of the IAUC's bibliographic database shows that work focused on the link between local climate knowledge and urban planning or development operations, although in vogue, is still marginal in this community. Only about 60 articles were published in international journals between 2015 and 2019, whereas the general scientific production in urban climatology amounted to about 3,000 articles over the same period.

3.2.2.2. *Increasing efforts to integrate climate and urban planning knowledge*

From a retrospective point of view, initial efforts focused on adapting the methods commonly used in the study of urban climatology (around modeling tools or cartography) to meet the needs of end users and facilitate communication toward decision-making. Since then, the range of resources mobilized has expanded, particularly from the human and social sciences, such as surveys and interviews or analysis of gray literature (institutional reports, white papers, reports published by non-governmental organizations).

For example, surveys have been used to qualify existing and future vulnerabilities in Istanbul (Onur and Tezer 2015). Živojinović and Wolfslehner (2015), meanwhile, used interviews to better understand the role of decision making for reforestation in Belgrade. This article assesses the perceptions of urban forest management actors toward climate change adaptation. Three types were identified: those who believe that climate variations are normal and that there is insufficient evidence of climate change; the "managers" who argue for more bottom-up initiatives and improved dialogue among stakeholders; and the "general awareness" advocates who believe that the issue of climate change should be governed by national organizations. Finally, Mason and Fragkias (2018) used *organization members' mental models* to study the potential of members of metropolitan planning organizations to influence public action and policy on climate change. The methods cited are a good example of how inter-disciplinarity can help us understand the lack of coordination and communication among stakeholders, the need for technical and scientific data at the local level and the need for specific regulations and funding.

The humanities and social sciences have also recently contributed to the debate on issues of equity, social vulnerability and the influence of non-climatic factors on vulnerability (Archer and Dodman 2015; Ryan 2015; Alonso and Renard 2020). Systems approaches using governance have been explored. Hatvani-Kovacs *et al.*

(2018) recommend integrated policy measures to increase the heat stress resilience of urban populations in Australian cities.

To facilitate transdisciplinarity, understood here as cooperation between scientific and non-scientific communities for the production of new action-oriented knowledge, and to reduce the negative impacts of heat waves on public health, urban infrastructure and services, we are witnessing in particular the development of socio-environmental systems modeling. For example, Kraucunas *et al.* (2015) propose a platform for integrated regional modeling and analysis (called PRIMA). It allows for the simulation of interactions between regional climate, hydrology, agriculture and land use, on the one hand, and human, socioeconomic and energy systems on the other hand, at decision-relevant scales. The platform can be customized according to specificities linked to the context (political, climatic, institutional).

In parallel, climate services are being developed in all their dimensions. The World Meteorological Organization recently published the "Guidance for Integrated Urban Hydrometeorological, Environmental and Climate Services"[3] (WMO 2019). These are the preliminary results of an action that aims to develop "Climate Services" at the urban scale. This action complements the four ongoing actions (agriculture, health, disaster prevention and water) of the Global Framework for Climate Services (GFCS), whose objective is to "strengthen the production, availability, delivery and application of science-based climate predictions and services". The GFCS was signed by 155 countries at the World Climate Conference-III in 2009.

Finally, cartographic tools such as "urban climatic maps" are used both for local-scale climate analysis (Ng and Ren 2015) and as intermediary objects (Hidalgo *et al.* 2019) that facilitate the communication and co-production of climate and urban planning knowledge, from environmental diagnostic to urban project.

3.2.2.3. *An update on mapping tools*

Historically, the German-speaking research community has been a pioneer in applied urban climatology studies, which have been systematized since the first half of the 20th century. Their techniques for mapping thermal distributions, ventilation corridors and air pollution concentrations, combining meteorological and climatic information with land use and topographical data have the advantage of meeting the needs of urban planning practices. The characterization and cartographic representation of the urban environment are currently an essential step in planning

3 https://library.wmo.int/doc_num.php?explnum_id=9903 [Accessed June 15, 2020].

and development exercises. They highlight the environmental amenities of the territory or serve to identify and locate environmental nuisances and risks in relation to elements as diverse as water, noise, vegetation, biodiversity or air quality.

In Germany, fine-scale climate mapping began in the 1970s with the "Climate Booklet for Urban Development" (1977) in Stuttgart. It continued with the "Klima Atlas" published in 1992. The methodology of the Stuttgart "Klima Atlas" was the subject of a German guideline, VDI3787 entitled "Climate and Air Pollution Maps for Cities and Regions", published in 1993 and updated in 2015. This guideline was published in German and English, which facilitated its circulation in Europe, Asia and South America for the most part.

In recent years, climate maps for urban planning have thus become a reference tool for the translation of microclimatic knowledge into planning guidelines and recommendations. Different names are used depending on the country and the scale of action, but in English-speaking literature the name Urban Climatic Maps (UC-Map) dominates today. Technical choices for the production of the maps depend on the local climatological characteristics, the available data on urban morphology and climate, and the scientific and cultural context of the actors (researchers and stakeholders) involved in the project. The climate data can be obtained by physical measurements (permanent meteorological network, field campaigns or remote sensing) or from a statistic or physical modeling using an atmospheric model. The management and representation of these data are generally done with the help of GIS.

This set of maps represents both a microclimatic diagnostic tool for the urban area and a potential lever for subsequent regulatory translation of the issues identified. This tool presents two levels of information:

– the first is the microclimatic diagnosis that gathers information on meteorology, land use, topography and vegetation. The interrelations between these elements and their effects on wind and thermal comfort are analyzed and evaluated spatially. This diagnosis is based on a set of maps, commonly called "analysis maps" (Figure 3.1), and crossed with planning elements (green and blue grid, zoning);

– the analysis maps lead to the identification of a certain number of zones at stake from a point of view that may be:

- thermal (typologies of urban areas according to their density or urban structure, thermoregulatory spaces such as water and vegetation, commercial and/or activity areas often highly mineralized and massively air-conditioned);

- aeraulics (predominant wind corridors, slope breeze generating areas, cool air producing areas – vegetation, water – or warm air producing areas – heavily air-conditioned – and areas that impede the wind);

– the second level of information is transcribed via a "recommendation map". It allows a translation of the general and practical strategic orientations in urban planning to improve the thermal and wind environments. It is derived from the analysis maps but also from the urbanistic and socio-cultural constraints.

Globally, in terms of the number of cities, the use of these maps remains low (Figure 3.2). There are even fewer cases where the translation of planning recommendations derived from climate diagnostics is actually applied at the operational level in land zoning, design codes or green space and water management.

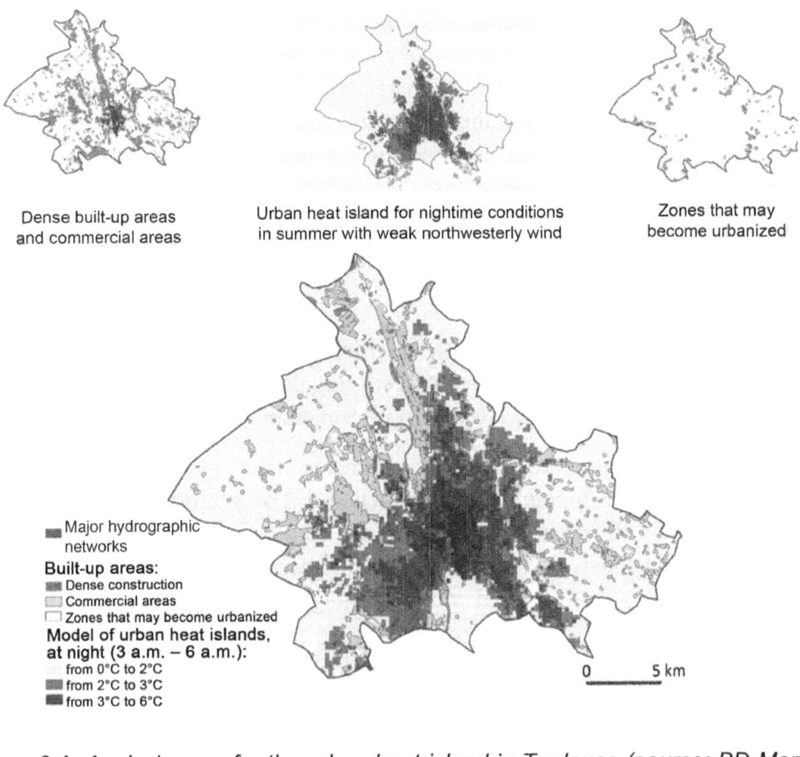

Figure 3.1. *Analysis map for the urban heat island in Toulouse (source: BD-Mapuce (2015), climate data (CNRM, LISST, 2018), Toulouse Métropole (2018), map design: N. Touati, J. Hidalgo, LISST (2020)). For a color version of this figure, see www.iste.co.uk/rebotier/risks.zip*

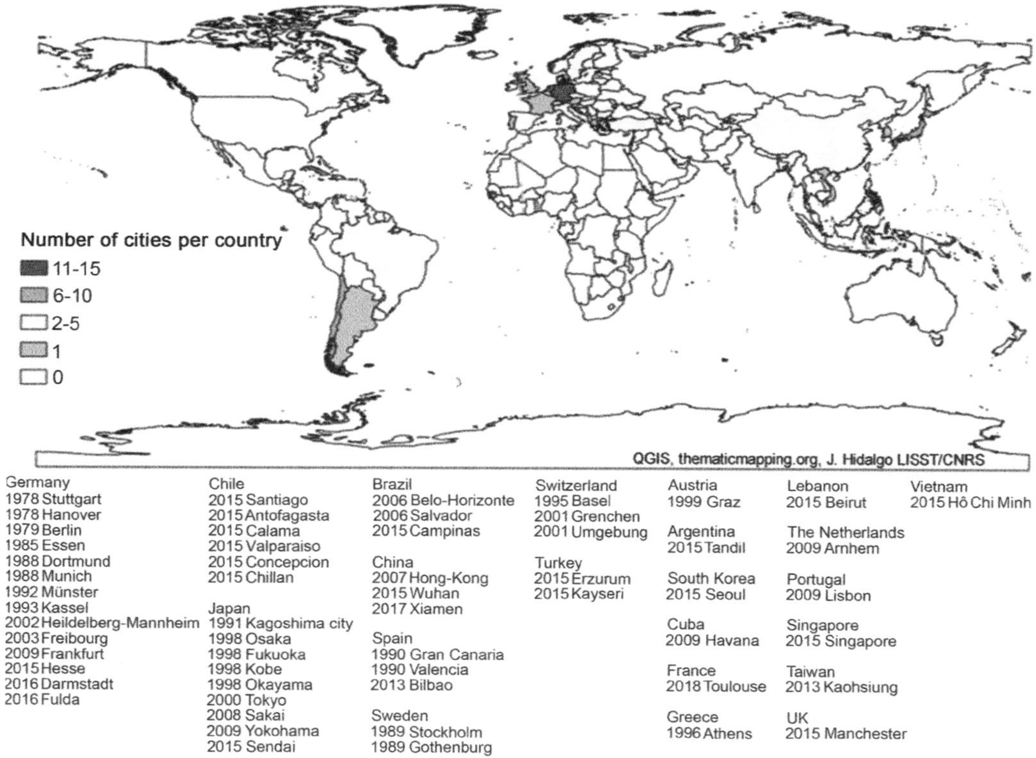

Figure 3.2. *Application of climate maps in urban planning (from Hidalgo* et al. *2019). The colors indicate the number of cities per country that have conducted studies drawing on this lineage of mapping work. The table corresponds to the date of the first published study for each city. For a color version of this figure, see www.iste.co.uk/rebotier/risks.zip*

3.2.3. *The difficulties of operationalizing knowledge about urban climate risk*

Given that the scientific basis and technical tools for local climate diagnosis exist, why is climate information successfully transferred to action in some places and not in others?

3.2.3.1. *The scale of intervention*

At the urban scale, climate governance is multi-level in many ways. It is organized through institutions that establish regulatory frameworks at the municipal, regional and national levels, but it also takes place through interventions designed and implemented by non-state actors, such as businesses, non-governmental organizations and civil society.

Aylett (2015) analyzed the institutional governance structures for climate adaptation put in place by local governments for 350 municipalities on five continents. The results show a rather cross-cutting or indirect integration of adaptation into long-term and sectoral territorial management plans, as adaptation does not fit neatly into institutional silos. Many strategic municipal sectors (such as those responsible for water, wastewater, health and urban regulation) remain on the margins of adaptation efforts. The level of comprehensiveness in climate-related policies is often correlated with linkages between researchers, NGOs (non-governmental organizations) and actors in the operational sphere (Lee and Painter 2015). In Australia, Graham and Mitchell (2016) analyze how partnerships between urban planners and frontier organizations influence adaptation planning. Disseminating knowledge and opening up the debate to actors and organizations that operate at the boundaries of the different social universes involved (decision-makers, technicians, residents) expands the scope of climate-related policies.

Without neglecting coordination with other national or regional scales, decentralization can strengthen the adaptive capacities of cities through the redistribution of budgetary, political and administrative competencies from centralized to local governments. This is pointed out by Moloney and Fanfgeld (2015) and Rumbach (2016), among others.

Finally, the "Global Research and Action Agenda on Cities and Climate Change Science" proposed by the academic community at the end of the IPCC (Intergovernmental Panel on Climate Change) Cities and Climate Change conference in Edmonton (2018) emphasizes three other key points that perfectly illustrate the complexity of taking climate issues into account at the local scale. These are the asymmetry between the governance capacity of cities in the South and those in the North, the different temporalities between desirable actions and the context of their evaluation, and the importance of interactions between spatial and temporal scales. For example, the effectiveness of actions at the local level depends on decisions taken at other spatial levels.

3.2.3.2. *The weight of disciplinary segmentation, between research and action*

Disciplinary segmentation is another difficulty on the road to operationalizing urban climate research. Some studies may have been commissioned by practitioners, notably for health reasons related to the air quality issue, but for the most part urban climatology projects are promoted and funded by producers of climate knowledge and data, namely academics or national meteorological services.

For example, Météo-France was at the origin of the first urbanized atmospheric simulation model in the early 2000s (Masson 2000) in order to improve its weather forecasting system. This development then allowed the initiation of modeling in urban climatology. At the beginning, the city was for atmospheric physicists an environment among others (forest, lake and ocean). Urban climate models were initially designed to calculate the impact of the built environment on weather variables. Over time and with the increase in computing power, they have become more refined in their representation, for example, for urban vegetation interacting with built elements, for water exchanges between the atmosphere, the surface and the ground, as well as for building energy. Demographic trends, urban expansion, or architectural evolution can also be taken into account in the modeling based on collaborations with economists, geographers, architects or urban planners (Masson *et al.* 2014).

Interdisciplinarity operates mainly through the integration of methods, models and data from disciplines that have historically invested in this field of research: atmospheric physicists, climate geographers and architects. At the margin, we find the more recent investment of researchers in social geography and urban planning, or in epidemiology. The practice of transdisciplinarity remains very recent due to the very strong weight of research in the proposal of projects. Nevertheless, it appears to be a clear trend in recent years.

A few studies have recently addressed issues of inter- and trans-disciplinarity. Groot *et al.* (2015) highlight through the case of Rotterdam the challenges that the relative newness of urban adaptation to climate change pose for the scientists and stakeholders involved. Sarzynski (2015) and Fagan-Watson and Burchell (2016) reveal that the role of public participation in such transdisciplinary integration efforts remains poorly theorized and understudied. In the first study, six forms of participation in urban adaptation are identified from the literature and the characteristics and use of each type are discussed. In the second study, the authors point out that regional and national policy documents that provide for action on heat waves in the UK tend to focus on emergency institutional responses and infrastructure development. In these documents, although communities are mentioned, they appear as passive beneficiaries of a top-down institutional resilience.

3.2.3.3. *The virtues of bottom-up processes: skills development and interdepartmental dissemination*

The legislative and regulatory injunctions that departments in charge of territorial planning are subject to can be tedious when added to the various environmental themes already addressed (biodiversity, air quality, noise). Nevertheless, it is observed that successful urban climate management experiences have two key ingredients (Lopez *et al.* 2015; Webb 2017): a local research team, taken to the extreme in Stuttgart with the presence of in-house urban climatologists, and a long process of collaboration and acculturation between researchers and practitioners. Progress in urban climate research does seem to be stimulated by international collaborations and networks as well as by national regulations. However, effective information transfer and implementation tends to be bottom-up. This is not to say that targets set by the national government are not an effective way to stimulate the development of environmental and climate codes at the local level (Heidrich *et al.* 2013; Olazabal *et al.* 2014; Reckien *et al.* 2014). Rather, there is evidence of a successful bottom-up, joint process (between researchers and practitioners) for the application of site-specific policies (sometimes beyond mere national or regional regulatory obligations), for their stabilization, and ultimately for their institutionalization over time. This fact must be taken into account by the national and international meteorological services that are currently designing future climate services at the urban scale.

In addition to effective sharing of information, tools and methods, internalization of expertise and cross-departmental dissemination at the city level are needed to ensure self-sufficiency in the field and to support the implementation of site-specific policies and regulations. Sustainable collaborations between researchers and practitioners allow for the transition from the study stage (through measurement or modeling campaigns to establish diagnoses) to the implementation stage, passing through the necessary stages of experimentation and formalization (via test sites and operations, the establishment of specific regulations and/or guidelines).

Experimentation is often used to test recommendations as well as to improve and adapt decision support systems. It also allows for the sensitization and information of departments other than the one in charge of implementing climate plans around concepts related to climate adaptation. The intelligibility of the information and messages that circulate seems essential, as does the presence of intermediaries who combine different skills and a certain hybridity of profiles: research/operational, urban planning/climatology (Lopez *et al.* 2015; Hidalgo *et al.* 2018).

3.3. Feedback of knowledge from the local to the global

It is true that putting climate change on the agenda is an "opportunity" to put urban climatology issues on the local agenda. But technical progress (evolution of modeling tools, production of databases and computing power) and experiments in the co-construction of knowledge at the local level have also enabled the urban climatology research community to contribute to the reflection and production of climate knowledge at the global level.

New developments include the WUDAPT urban database, which provides the urban data needed for climate modeling for a large number of cities. Long-term urban climate modeling has not been included in previous IPCC exercises because practically no regional climate model has a physical representation of the urban surface. However, the new generation of regional climate models with very high resolution coupled with urban models will make it possible to link research communities in urban meteorology and climate change modeling.

Nevertheless, there are some pioneering studies. Daniel *et al.* (2019) use a regional "urbanized" model and quantify the urban climate in France over 30 years. By focusing on eight large cities, they show that they induce regional warming well beyond the boundaries of urban areas. Garuma *et al.* (2018) use similar simulations on the eastern North American continent. They find that in pessimistic scenarios (RCP 8.5), urban heat island intensity increases by 0.5°C on average. However, these climate simulations are still too coarse (25 km of horizontal resolution) to be able to correctly resolve the urban heat island phenomenon. Other methods exist, however, that use high-resolution simulation models in conjunction with statistical processing, making the processing lighter from a computational point of view (Kusaka *et al.* 2012; Hamdi *et al.* 2014, 2016).

The IPCC has recognized the leading role of cities in responding to climate change and has proposed that the seventh assessment cycle include a "Special Report on Climate Change and Cities". To stimulate knowledge generation and exchange on this topic, the IPCC organized a Cities and Climate Change Science Conference – CitiesIPCC – in Edmonton in March 2018, supported by various international networks and associations (C40, Cities Alliance, ICLEI-Local Governments for Sustainability, Future Earth, SDSN-Sustainable Development

Solutions Network, UCLG-United Cities and Local Governments, UN-Habitat, UN-Environment, and WCRP-World Climate Research Program). The event brought together representatives from academia, scientific institutions, IPCC experts, national, regional and local government representatives and practitioners.

The synthesis of the contributions led to the drafting of a "Global Research and Action Agenda on Cities and Climate Change Science"[4]. This document aims to guide future research to make it useful for "effective policy development for climate action in cities".

Only a few sessions have been organized in Edmonton by IAUC members, and very few urban climatology studies are cited in the chapter on urban areas in the IPCC 1.5°C report. The search for a certain uniformity of analysis methods explains much of this parsimony. But this uniformity runs counter to the richness of the spatial and temporal descriptions needed for urban climatology studies. A major scientific challenge is therefore emerging in the near future: to advance the production of large-scale knowledge that feeds into global agendas, without giving up high spatial resolution or the specificities that characterize local climate issues.

3.4. References

Akbari, H. and Kolokotsa, D. (2016). Three decades of urban heat islands and mitigation technologies research. *Energy and Buildings*, 133, 834–842.

Aleksandrowicz, O., Vuckovic, M., Kiesel, K., Mahdavi, A. (2017). Current trends in urban heat island mitigation research: Observations based on a comprehensive research repository. *Urban Climate*, 21, 1–26.

Alonso, L. and Renard, F. (2020). A comparative study of the physiological and socio-economic vulnerabilities to heat waves of the population of the metropolis of Lyon (France) in a climate change contex. *International Journal of Environmental Research and Public Health*, 17(3), 1004.

Archer, D. and Dodman, D. (2015). Making capacity building critical: Power and justice in building urban climate resilience in Indonesia and Thailand. *Urban Climate*, 14(part 1), 68–78.

Aylett, A. (2015). Institutionalizing the urban governance of climate change adaptation: Results of an international survey. *Urban Climate*, 14(part 1), 4–16.

Barlow, J. (2014). Progress in observing and modelling the urban boundary layer. *Urban Climate*, 10(2), 216–240.

4 www.wcrp-climate.org/WCRP-publications/2019/GRAA-Cities-and-Climate-Change-Science-Full.pdf [Accessed June 15, 2020].

Baumüller, J. (2005). Stuttgart21, Amt für Umweltschutz. Stuttgart, version 4.

Bulkeley, H. and Betsill, H. (2003). *Cities and Climate Change*. Routledge, London.

Chapman, L. (2015). Urban meteorological networks: An urban climatologists panacea? *ICUC-9*, Toulouse, France.

Ching, J., Mills, G., Bechtel, B., See, L., Feddema, J., Wang, X., Ren, C., Brousse, O., Martilli, A., Neophytou, M., Mouzourides, P., Stewart, I., Hanna, A., Ng, E., Foley, M., Alexander, P., Aliaga, D., Niyogi, D., Shreevastava, A., Bhalachandram, S., Masson, V., Hidalgo, J., Fung, J., de Fatima Andrade, M., Baklanov, A., Wei Dai, D., Milcinski, G., Demuzere, M., Brunsell, N., Pesaresi, M., Miao, S., Chen, F. (2018). World Urban Database and Access Portal Tools (WU-DAPT), an urban weather, climate and environmental modeling infrastructure for the Anthropocene. *Bulletin of the American Meteorological Society*, 99(9), 1907–1924.

Dabberdt, W., Frederick, G., Hardesty, R., Lee, W.C., Underwood, K. (2004). Advances in meteorological instrumentation for air quality and emergency response. *Meteorol. Atmos. Phys.*, 87, 57–88 [Online]. Available at: https://doi.org/10.1007/s00703-003-0061-8 [Accessed 15 June 2020].

Daniel, M., Lemonsu, A., Déqué, M., Somot, S., Alias, A., Masson, A. (2019). Benefits of explicit urban parameterization in regional climate modeling to study climate and city interactions. *Climate Dynamics*, 52(5–6), 2745–2764 [Online]. Available at: https://doi.org/10.1007/s00382-018-4289-x [Accessed 5 June 2020].

Fagan-Watson, B. and Burchell, K. (2016). Heatwave planning: Community involvement in co-producing resilience. *Building Research & Information*, 44(7), 754–763.

Garuma, G.F. (2018). Review of urban surface parameterizations for numerical climate model. *Urban Climate*, 24, 830–851 [Online]. Available at: https://doi.org/10.1016/j.uclim.2017.10.006 [Accessed 15 June 2020].

Goldreich, Y. (2003). *The Climate of Israel: Observation, Research and Application*. Springer, New York.

Grafakos, S., Viero, G., Reckien, D., Trigg, K., Viguie, V., Sudmant, A., Graves, C., Foley, A., Heidrich, O., Mirailles, J., Carter, J., Chang, L., Nador, C., Liseri, M., Chelleri, L., Orru, H., Orru, K., Aelenei, R., Bilska, A., Pfeiffer, B., Lepetit, Q., Church, J., Landauer, M., Gouldson, A., Dawson, R. (2020). Integration of mitigation and adaptation in urban climate change action plans in Europe: A systematic assessment. *Renewable and Sustainable Energy Reviews*, 121(109623) [Online]. Available at: https://www.sciencedirect.com/science/article/pii/S1364032119308305.

Graham, A. and Mitchell, C.L. (2016). The role of boundary organizations in climate change adaptation from the perspective of municipal practitioners. *Climatic Change*, 139(3), 381–395.

Groot, A.M.E., Bosch, P.R., Buijs, S., Jacobs, C.M.J., Moors, E.J. (2015). Integration in urban climate adaptation: Lessons from Rotterdam on integration between scientific disciplines and integration between scientific and stakeholder knowledge. *Building and Environment*, 83, 177–188.

Hamdi, R., Van de Vyver, H., De Troch, R., Termonia, P. (2014). Assessment of three dynamical urban climate downscaling methods: Brussels's future urban heat island under an A1B emission scenario. *International Journal of Climatology*, 34(4), 978–999.

Hamdi, R., Duchêne, F., Berckmans, J., Delcloo, A., Vanpoucke, C., Termonia, P. (2016). Evolution of urban heat wave intensity for the Brussels Capital Region in the ARPEGE-Climat A1B scenario. *Urban Climate*, 17, 176–195.

Hatvani-Kovacs, G., Bush, J., Sharifi, E., Boland, J. (2018). Policy recommendations to increase urban heat stress resilience. *Urban Climate*, 25, 51–63.

Hebbert, M. and Mackillop, F. (2013). Urban climatology applied to urban planning: A postwar knowledge circulation failure. *International Journal of Urban and Regional Research*, 37, 1542–1558.

Heidrich, O., Dawson, R.J., Reckien, D., Walsh, C.L. (2013). Assessment of the climate preparedness of 30 urban areas in the UK. *Climatic Change*, 120, 771–784.

Hidalgo, J., Haouès-Jouve, S., Medhbi, Z., Bretagne, G., Guilouhfi, J., Lau, K., Moure, T., Ng, E., Van-Hammers, H. (2018). Initiating climatic awareness in urban planning practices through participatory action research. *Cities and Climate Change Science Conference*, Edmonton, Canada.

Hidalgo, J., Lemonsu, A., Masson, V. (2019). Between progress and obstacles on urban climate interdisciplinary studies and knowledge transfer to society. *Annals of the New York Academy of Sciences*, 1436(1), 5–18.

Hintz, M.J., Luederitz, C., Lang, D.J., Wehrden, H. (2018). Facing the heat: A systematic literature review exploring the transferability of solutions to cope with urban heat waves. *Urban Climate*, 24, 714–727.

Horsfall, T.C. (1904). *Improvement of the Dwellings and Surroundings of the People: The Example of Germany*. Manchester University Press, Manchester.

Kanda, M. (2006). Progress in the scale modeling of urban climate: Review. *Theor. Appl. Climatol.*, 84, 23–33 [Online]. Available at: https://doi.org/10.1007/s00704-005-0141-4 [Accessed 15 June 2020].

Kassner, C. (eds) (1910). *Die meteorologische Grundlagen des Städtebaus [The Meteorological Basics of City Planning]*. Ernst und Sohn, Berlin.

Kratzer, A. (1937). *Das Stadtklima [City Climate]*. Friedr. Vieweg & Sohn, Braunschweig.

Kraucunas, I., Clarke, L., Dirks, J., Hathaway, J., Hejazi, M., Hibbard, K., Huang, M., Jin, C., Kintner Meyer, M., Van Dam, K., Leung, R., Li, H.Y., Moss, R., Peterson, M., Rice, J., Scott, M., Thomson, A., Voisin, N., West, T. (2015). Investigating the nexus of climate, energy, water, and land at decision-relevant scales: The Platform for Regional Integrated Modeling and Analysis (PRIMA). *Climatic Change*, 129(3–4), 573–588.

Kusaka, H. (2008). Recent progress on urban climate study in Japan. *Geographical Review of Japan*, 81(5), 361–374 [Online]. Available at: http://doi.org/10.4157/grj.81.361 [Accessed 15 June 2020].

Kusaka, H., Hara, M., Takane, Y. (2012). Urban climate projection by the WRF model at 3-km horizontal grid increment: Dynamical downscaling and predicting heat stress in the 2070's August for Tokyo, Osaka, and Nagoya Metropolises. *Journal of the Meteorological Society of Japan*, 90B, 47–63 [Online]. Available at: https://doi.org/10.2151/jmsj.2012-B04 [Accessed 15 June 2020].

Lambert-Habib, M.L., Hidalgo, J., Fedele, C., Lemonsu, A., Bernard, C. (2013). How is climatic adaptation of cities taken into account by legal tools? Focus on the introduction of water and vegetation in the French town planning documents. *Urban Climate*, 4, 16–34.

Le Roy, B., Lemonsu, A., Schoetter, R. (2021). A statistical-dynamical downscaling methodology for the urban heat island applied to the EURO-CORDEX ensemble. *Climate Dynamics*, 56, 2487–2508.

Lee, T. and Painter, M. (2015). Comprehensive local climate policy: The role of urban governance. *Urban Climate*, 14(part 4), 566–577.

Lopez, X., Haouès-Jouve, S., Hidalgo, J. (2015). Integration of urban climate issues in urban planning: Reflections on which are the keys of success. *ICUC9*, Toulouse, France.

Lu, C., Huang, G., Wang, X., Liu, L. (2021). Ensemble projection of city-level temperature extremes with stepwise cluster analysis. *Climate Dynamics*, 56, 3313–3335.

Mason, S.G. and Fragkias, M. (2018). Metropolitan planning organizations and climate change action. *Urban Climate*, 25, 37–50.

Masson, V. (2000). A physically-based scheme for the urban energy budget in atmospheric models. *Boundary-Layer Meteorol.*, 94, 357–397.

Masson, V. (2006). Urban surface modelling and the meso-scale impact of cities. *Theoretical and Applied Climatology*, 84, 35–45.

Masson, V., Marchadier, C., Adolphe, L., Aguejdad, R., Avner, P., Bonhomme, M., Bretagne, G., Briottet, X., Bueno, B., de Munck, C., Doukari, O., Hallegatte, S., Hidalgo, J., Houet, T., Le Bras, J., Lemonsu, A., Long, N., Moine, M.P., Morel, T., Nolorgues, L., Pigeon, G., Salagnac, J.L., Zibouche, K. (2014). Adapting cities to climate change: A systemic modelling approach. *Urban Climate*, 10, 407–429.

Moloney, S. and Fanfgeld, H. (2015). Emergent processes of adaptive capacity building: Local government climate change alliances and networks in Melbourne. *Urban Climate*, 14(part 1), 30–40.

Ng, E. and Ren, C. (2018). China's adaptation to climate and urban climatic changes: A critical review. *Urban Climate*, 23, 352–372.

Oke, T.R., Mills, G., Christen, A., Voogt, J. (2017). *Urban Climate*. Cambridge University Press, Cambridge.

Olazabal, O., De Gregorio Hurtado, S., Olazabal, E., Pietrapertosa, F., Salvia, M., Geneletti, D., D'Alonzo, V., Feliú, E., Di Leo, S., Reckien, D. (2014). How are Italian and Spanish cities tackling climate change? A local comparative study. *BC3 Working paper series*, 03.

Onur, A.C. and Tezer, A. (2015). Ecosystem services based spatial planning decision making for adaptation to climate changes. *Habitat International*, 47, 267–278.

Pisello, A.L. (2017). State of the art on the development of cool coatings for buildings and cities. *Solar Energy*, 144, 660–680.

Reckien, D., Flacke, J., Dawson, R., Heidrich, O., Olazabal, M., Foley, A., Hamann, J.P., Orru, H., Salvia, M., De Gregorio Hurtado, S., Geneletti, D., Pietrapertosa, F. (2014). Climate change response in Europe: What's the reality? Analysis of adaptation and mitigation plans from 200 urban areas in 11 countries. *Climatic Change*, 122, 331–340.

Reckien, D., Salvia, M., Pietrapertosa, F., Simoes, S., Olazabal, M., Hurtado, S.D.G., Geneletti, D., Lorencová, E.K., D'Alonzo, V., Krook-Riekkola, A., Fokaides, P., Ioannou, B., Foley, A., Orru, H., Orru, K., Wejs, A., Flacke, J., Church, J., Feliu, E., Vasilie, S., Nador, C., Matosović, M., Flamos, A., Spyridaki, N.A., Balzan, M., Fülöp, O., Grafakos, S., Paspaldzhiev, I., Heidrich, O. (2019). Dedicated versus mainstreaming approaches in local climate plans in Europe. *Renewable and Sustainable Energy Reviews*, 112, 948–959.

Rumbach, A. (2016). Decentralization and small cities: Towards more effective urban disaster governance? *Habitat International*, 52, 35–42.

Ryan, D. (2015). From commitment to action: A literature review on climate policy implementation at city level. *Climatic Change*, 131(4), 519–529.

Sarzynski, A. (2015). Public participation, civic capacity, and climate change adaptation in cities. *Urban Climate*, 14(part 1), 52–67.

Shooshtarian, S., Rajagopalan, P., Sagoo, A. (2018). A comprehensive review of thermal adaptive strategies in outdoor spaces. *Sustainable Cities and Society*, 41, 647–665.

Stewart, I.D. and Oke, T.R. (2012). Local climate zones for urban temperature studies. *Bull. Am. Meteorol. Soc.*, 93, 1879–1900.

Stock, P. (1992). Climatic classification of town areas. In *Planning Applications of Urban and Building Climatology*, Höschele, K. (ed.). Institut für Meteorologie und Klimaforschung – Universität Karlsruhe, Karlsruhe.

Stock, P. and Beckröge, W. (1985). Klimaanalyse Stadt Essen. KVR, PO15, Planungshefte Ruhrgebiet, Essen.

Tanaka, T., Ogasawara, T., Koshi, H., Yoshida, S. (2009). Urban environmental climate maps for supporting urban-planning related work of local governments in Japan: Case studies of Yokohama and Sakai. *ICUC7*, Yokohama, Japan.

Thibaud, J.P. (2018). Vers une écologie ambiante de l'urbain. *Environnement Urbain*, 13 [Online]. Available at: www.erudit.org/fr/revues/eue/2018-v13-eue04046/1052702ar/ [Accessed 5 October 2020].

UCCRN (2011). *Climate Change and Cities (First Assessment Report of the Urban Climate Change Research Network)*. Cambridge University Press, Cambridge.

UN (2011). *Cities and Climate Change: UN-Habitat Global Report on Human Settlements*. Earthscan, London.

Webb, B. (2017). The use of urban climatology in local climate change strategies: A comparative perspective. *International Planning Studies*, 22(2), 68–84.

WMO (2017). *Guide to Meteorological Instruments and Methods of Observation.* Part II, Chapters 9 and 10 [Online]. Available at: www.weather.gov/media/epz/mesonet/CWOP-WMO8.pdf [Accessed 15 June 2020]

WMO (2019). *Guidance on Integrated Urban Hydrometeorological, Climate and Environment Services. Volume I: Concept and Methodology*. WMO-No. 1234.

Živojinović, I. and Wolfslehner, B. (2015). Perceptions of urban forestry stakeholders about climate change adaptation – A Q-method application in Serbia. *Urban Forestry & Urban Greening*, 14(4), 1079–1087.

PART 2

Recompositions for the Study and Management of Risks?

4

Permanence and Specificities of Risks and Their Management in the Anthropocene Era

Patrick PIGEON
MÉDIATIONS, Université Savoie Mont Blanc, Chambéry, France

4.1. The Anthropocene, paradox of a new era?

4.1.1. *The multiplication of research programs and prevention policies in the age of the Anthropocene*

The questions that concern the Anthropocene imply links between research and action. Research identifies and interprets the current evolutions of the planet Earth presented as highly problematic on the epistemological, ethical or economic levels, in their interactions with the evolutions of humanity. This research justifies an injunction to political action. The modalities of these political actions are notably presented and discussed by disaster prevention programs. The study of the relationship between humanity and planet Earth is by no means new. Ecology, as an academic discipline, was born of the desire to better understand the interactions between human societies and certain ecosystems, with the hope of proposing solutions to the problems then identified, through political measures (Drouin 1993, p. 87, in particular with the work of Karl Möbius). The discipline of geography also contributed to this project, with famous precursors such as Humboldt. In France, Max Sorre has developed a geographical thought that is much marked by an ecosystemic approach and which is nonetheless applied. The "pathogen complex" helped to found the discipline of medical geography (Simon 2016). Under these conditions, what would the notion of the Anthropocene bring to ecology and geography?

Despite the minimal consensus that the neologism was launched at the very beginning of the 21st century by an atmospheric chemist (Fressoz and Bonneuil 2013; Scott 2017; Magny 2019), the evidence and chronological markers for identifying and justifying the academic notion of the Anthropocene remain debated (see the Introduction to this book). The controversy surrounding the Anthropocene seems to draw attention to processes presented as radically new on a global scale. They call into question the distinction between the long duration, that of nature, geological time, and the time of human societies, historical time, etc. These processes, such as those associated with changes in the climate or the biosphere, are also presented as heralding disasters, for example, in connection with heat waves and urban heat islands (see Chapter 3), pandemics, storms or significantly recomposed coastal risks (see Chapter 1). Understood in this way, the notion of the Anthropocene reconsiders studies that are already old, dating back to the 19th century, in ecology. It calls into question the relevance of the disciplinary divisions between human sciences and earth sciences in order to better understand and manage the transformative capacities of humanity both on a longer time scale and on a worldwide scale.

Indeed, the bibliography agrees that the Anthropocene describes humanity's capacity to transform planet Earth, in ways that partly unintentionally increase the risk of disasters. The political field of risk management then joins the academic field in the face of such challenges. The Anthropocene can, for example, feed into *collapse studies*, which were recently disseminated by the American geographer Jared Diamond (2005), while being contested (Scott 2017). In principle at least, the notion also has the potential to bring together groups of researchers who have difficulty exchanging ideas: those, mainly on the side of the humanities and social sciences, who work on disaster risk reduction, and those, mainly on the side of the Earth sciences, who identify, among other things, the marks of climate change linked to anthropization (Pigeon and Rebotier 2016).

Even though it is the subject of debate, recognition of the Anthropocene has several major consequences that are, in substance, very much related:

– the notion of the Anthropocene implies more and more approaches, among which we find policies for risk prevention or adaptation. The more human societies are responsible for their own direct or indirect actions on planet Earth, the more policies are justified, at least logically. This is what we will illustrate below with the evolution of mountain land restoration policies. What is of human origin, even partially, can be corrected by humans, especially if it is a matter of disaster prevention. The ethical argument seems to reinforce the logical argument;

– the tendency to multiply these policies increases the recognition of human responsibilities (Jonas 1979), instead of invoking major causes that would escape humanity, because they are natural. The neologism of the anthropocene insists on this now dominant responsibility, which is that of humanity, and which did not appear in the other, older neologism, that of ecology;

– but more and more policies are drawing more and more attention to the limits of policies, calling them into question on ever shorter time scales (Revault d'Allonnes 1999). It is also time for reflexivity, the potential capacity to reconsider the results of a decision or an action (Beck 1992). The recognition of humanity's capacity to transform the climate and to contribute to the involuntary creation of future disasters also contributes to this reflexivity;

– the notion of the Anthropocene requires more interdisciplinary research. The Anthropocene invites us even more to contest analyses in terms of human/nature duality. The notion calls for hybrid analyses, more adapted to the understanding of supposedly increased forms of complexity. Morin's (1977, 2014) reflections show that this trend is not new. Disciplinary research remains useful, but necessarily limited when it comes to accounting for hybrid phenomena. This is even more the case at the time of the Anthropocene;

– policy makers and research program designers are, thus, confronted with the challenges posed by the Anthropocene: what can be done to push back the recognized limits of sectoral actions and thinking, which are nonetheless necessary (Pigeon *et al.* 2018)? The Anthropocene, by its identification, seems to justify not only interdisciplinarity but also transdisciplinarity: policy makers, managers, research program designers and researchers are encouraged to develop more integrated research and policies. These pose formidable epistemological problems. In addition, there are problems of social justice, because, faced with environmental changes concerning the planet, not everyone is equal (as discussed in Chapters 6 and 8).

The multiplication of risk prevention policies, as well as research programs on these issues, is thus accompanied by major problems, without it being possible to find a definitive answer. This trend seems all the more entrenched during this Anthropocene moment. It is presented as a necessity that is as justifiable as it is paradoxical, notably by the international institutions that relay it, promote it and disseminate it (WEF 2017; WMO 2017; WB 2019). The Anthropocene moment partly calls into question pre-existing ways of thinking and acting, while at the same time immediately highlighting the limits of the solutions advocated, since one path or initiative is followed by several others. This situation seems to fuel a profusion of recent theses, increasingly conducted in partnership with managers (Provitolo 2002; Combe 2007; Gérin 2010; Carlier 2019).

4.1.2. *A new word for old evils?*

For all that, the intensification of knowledge and management efforts at the time of the Anthropocene cannot constitute a fundamental break, despite what international institutions and a part of the research community put forward. This latter inter-prediction is contested by recent summaries on the issue (Fressoz and Bonneuil 2013; Scott 2017), or by a number of previous works on the relations between urbanization and environmental change (Pigeon 1994, 2012a). The same is true for disciplinary division: its mutilating character has long been denounced by philosophers like Bergson (2014) and Morin (1977).

Under these conditions, posing the same problems in the light of a new word, this neologism called the Anthropocene, does not seem to call into question the foundations of academic research or modes of political decision making. The problems addressed through the prism of the Anthropocene seem to remain epistemological, and as inherent to humanity, over the very long philosophical period. If this is the case, how can we understand the recent appearance of the Anthropocene? For what, and above all, for whom, can this notion be useful? Why also such a multiplication of disaster prevention policies, while their dominant official presentation regularly insists on their limits, even their failures (Pigeon and Rebotier 2016)?

We propose answers to these questions through an analysis of policies related to the prevention of disaster risks that have existed for more than a century and a half in France: the restoration of mountainous terrain, which was supposed to respond to the misuse of the environment, the cause of flooding. The reconstruction of this evolution necessarily draws attention to the places where these policies are made and recomposed, which are the metropolises.

We deduce that disaster risk prevention policies, whether in the context of the Anthropocene moment or not, are also geopolitical. A multi-risk approach is, therefore, justified. Past policy experiences allow us to identify the use of prevention or post-disaster reconstruction policies for purposes other than the prevention of one type of risk. Consider the Anthropocene multiplies the justifications for these policies, and simultaneously their instrumentalization, for other purposes by a wide variety of actors. This reading makes it possible to argue that, as it stands, the recognition of the Anthropocene is not truly disruptive. It is not a profoundly transformative and innovative break with the dominant structures of thought and action. Rather, it is a continuation of what already exists. The analysis of disaster risk prevention policies linked to mountain land restoration policies makes it possible to defend this interpretation. This does not mean that these policies do not contribute to preventing certain disasters.

We find the same interpretation in analysis of contemporary European policies that seek to reduce state intervention in prevention policies. The trend is increasingly apparent in European research programs and associated theses that aim to influence prevention policies. Some European research programs emphasize the local scale and challenge the relevance of state or supralocal aid, as we will see in a few examples. It is at the local scale that the most "effective" decisions are supposed to be taken, allowing for the prevention of disasters, whether or not they are linked to the Anthropocene... Despite the effort to stand back critically, this promotion can also be found on the side of managers who defend "the revolution of proximity" at a time when "we have dramatically entered the era of the Anthropocene" (Farinelli 2015, p. 22). The tendency to value the local scale of management is also present in the work that Cerema engineers are conducting, on the search for an "applied resilience of territories" (Cerema 2019). However, this preference for action at local level is anything but neutral, anything but new, and highly ideologically conditioned. It even poses important rationality problems, primarily with regard to the classic definitions and understandings of what a disaster is. Indeed, a disaster is an event whose intensity of damage exceeds local management capacities. Strengthening local management capacities in order to avoid a disaster is, therefore, defensible and even logical. But hoping to rely solely on this strengthening of local capacities becomes illogical: among other things, this would imply investments that would exceed local budgetary and financial capacities. In concrete terms, can one argue that strengthening local resources alone would be sufficient to prevent a disaster? In France, the small size of the communes is enough to show the limits of the argument. The same is true for other countries, even those with a federal structure like the United States, as Comby's (2015) thesis has shown.

The continued instrumentalization of prevention policies, or the greater emphasis on the local under the guise of pragmatism, once again draw attention to the fact that, in the age of the Anthropocene, very old trends are being dressed up in new clothes. The dominant ways of thinking and, above all, of making political decisions are maintained. We defend this point with an analysis of mountain land restoration policies and their evolution.

4.2. Restoration of mountain terrain: recompositions of policies centered on the correction of hazards

4.2.1. *The history of the RMT laws: fundamental similarities with the justifications of the Anthropocene*

One may be surprised to see the global issue of the Anthropocene addressed from the perspective of mountain land restoration policies in France, which seem narrowly and locally specialized. Explicit links do exist, however: restoration of mountain terrain (RMT) policies now incorporate the notion of the Anthropocene to justify the evolution of the risk prevention measures they advocate. But this justification system is not new. The history of RMT policies illustrates, well before the work of Beck (1992), the tendency to reflexivity. And well before the reflections of Revault D'Allonnes (1999), the history of RMT laws also illustrates the paradox that the philosopher exposes: a risk prevention approach is necessary so that we can hope to correct the harmful consequences of previous actions, despite the irreducible limits of any preventive action. Also, well before the Anthropocene, the laws of RMT display the misuse associated with what is presented as a disruption of the environment to justify policies that, in the end, are above all turned toward work and the search for hazard control.

The laws of mountain land restoration, mainly those of July 28, 1860 and April 4, 1882, were officially aimed at preventing the risks of disasters associated with torrents, avalanches and landslides, but also with torrential rivers or large rivers, downstream. They postulated that these risks were largely increased by the demographic pressure on the upstream parts of torrential river basins, requiring "land restoration". This implied policies of major works, in particular reforestation and torrential correction.

In spite of these justifications, the publications on the history of mountain land restoration laws all insist on their slow and difficult genesis, as well as on their questioning. After the failure of the 1843 attempt in the face of what Brugnot and Cassayre (2003) call "the agro-pastoral lobby" (challenged by the RMT approach), the law of 1860 and above all that of April 4, 1882 laid the foundations for the restoration of mountain land. The period is presented as a sort of golden age by RMT managers (Yves Cassayre was director of the RMT service of Haute-Savoie during the 1980s). The impressive sum constituted by Mougin's book (1914) seems to impose such a judgment. More than a thousand pages detail the work carried out under the aegis of this policy, on the scale of the torrential river catchment areas. But the disasters linked to floods (such as those of Lyon in 1841 and 1856) (Combe

2007), which seemed to justify the appearance of the main RMT laws, were just as much used to question them.

4.2.2. *From the limits of RMT to risk management in the Anthropocene era*

Indeed, both reforestation and correction works failed to prevent further disasters, such as those in Val d'Isère (February 10, 1970) and the Plateau d'Assy (April 16, 1970), both in the northern French Alps. The result was a reorganization of RMT policies in 1980, which oriented preventive policies toward more regulation of urbanization. This was made possible by the PERNP (*plans d'expositions aux risques naturels prévisibles*) (plans for exposure to predictable natural risks) born of the Tazieff law (July 13, 1982) and its main application decree (in 1984) (Pigeon 1991). Since 1982–1984, the PERNPs, which became the risk prevention plans in 1995, have undergone numerous transformations. They are documented by a large number of theses on these subjects, notably those of Gérin (2010) and Carlier (2019). Consideration of European regulations contributed to this, particularly with the Flood Directive (from 2007).

The evolution of policies was thus driven by disasters that seemed to draw attention to the limits of pre-existing measures. In concrete terms, this was reflected in the multiplication of the tools and skills used: the forestry engineer (1860) also became a civil engineer, with the correction of torrents and torrential rivers, landslides and avalanches (1882). To this is added the competence of the cartographer involved in the negotiation of risk maps, on behalf of the State (from 1984). Despite the disengagement of the RMT services from this risk management activity involving the control of urbanization since 2005 (Arnould *et al.* 2016), the forestry engineer "finds himself an expert in the assessment of hazards and their consideration in land use planning (urban planning in particular) and in supporting crisis management" (Arnould *et al.* 2016, p. 28). In addition, today, there are injunctions to take into account the transformations of risks associated with climate change:

> Without resorting to catastrophism, the possibility of such events should be taken into account and it should be ensured that the zonings and rules of the PPRN leave the possibility of technically and economically feasible solutions for preventing these risks, should they occur and increase. (Arnould *et al.* 2016, p. 35)

RMT policies are thus led to think about disaster risk prevention in terms of the Anthropocene: one of the main areas of work is climate change.

This is a general evolution that goes beyond mountain environments. It can be found, for example, in research on the transformation and development of rivers in the Anthropocene era and on the associated policies. David (2016, p. 30) thus introduces "the response of societies to change" marked by "the era of restoration":

> Faced with the observation of a deterioration of environmental conditions, a desire to restore watercourses gradually emerged in the 1980s, on the scale of Western countries and particularly in Europe. This will is now widely assumed and put forward by the European Parliament, through the Water Framework Directive (WFD 2000). At the same time, it motivates numerous scientific studies, which have highlighted the benefit of a good knowledge of the historical trajectory of the systems to be restored in the success of the restoration processes.

The multiplication and complexity of policies is a fundamental trend. It is justified by the need to reconsider decisions previously made based in part on academic work. Clearly, notions such as restoration are still relevant, albeit debated (David 2016). Recognizing both the limitations of policies and the frameworks of thought that influence those policies is not enough to dismiss them permanently. There is something of a fatality in the inability to radically reconsider the policies to question the orders that underpin them. To understand such an ambiguous record, it is necessary to go back to the decision-making and financing processes, which have one thing in common: they are associated with capitals and metropolises.

4.3. Knowing, telling and managing risks: levers and attributes of metropolitan power

4.3.1. *Metropolises and disaster risk prevention: the weight of Paris, Brussels, Strasbourg and Geneva*

Metropolises are the dynamic crucibles where existing policies and research programs are conceived and challenged, and which are associated with legal norms. The etymology of the metropolis – *meter-polis*, the mother city – announces the supposedly innovative character of metropolises (Bonnet 1994; Reghezza 2007). It is similar to the etymology, this time Latin, of the capital, which refers to the thinking and directing head.

The history of the RMT laws confirms this: they were conceived and promulgated by Parliament, so in Paris. Later, the European legislative evolution, as well as the conception of European research programs, involved Brussels and Strasbourg. One of the official objectives of the laws associated partially or totally with the restoration of mountainous terrain was and remains to protect the main cities of France, such as Lyon (Bravard 1987; Combe 2007). The Flood Directive (2007) carries a similar message: it is from the decision-making centers of the European Union that some of the orientations of disaster risk prevention policies, and the associated research programs, are designed, even in part (Menoni and Margottini 2011; Aitsi-Selmi *et al.* 2016). Equally noteworthy is the influence of Geneva, where the United Nations Office for Disaster Risk Reduction (UNDRR) is located. The UN institutions are at the origin of the Sendai Framework for Action (Weichselgartner and Pigeon 2015), which is also taken up by the French Association for the Prevention of Natural Disasters, or the *Mission des risques naturels*, both of which are located in Paris. These places of decision and influence are also places where many lobbies are concentrated.

4.3.2. *The weight of central power players in defining initiatives*

The policies and programs that emphasize environmental approaches to risk management are in fact linked to many issues other than those associated with the environment. This is a structural, centuries-old trend: here again, the notion of the Anthropocene does not bring about a fundamental break with the RMT. Risk is always a matter of politics, and the Anthropocene moment is no exception.

The first RMT laws manifest, for example, the political will to control the mountain margins of the country, which are seen as borders. This was particularly the case for Savoy, which became part of France in 1860. The trip of Napoleon III to Savoy, the same year, remains attached to the symbol of the sovereign who defends a recently annexed territory, and who, just as much, submits it. Napoleon III's trip was hindered by the flooding of the Foron River in Scionzier in September 1860 (Mougin 1914), and then on the way up to Chamonix, without making it impossible. A memory of it remains in the commune of Scionzier (Haute-Savoie). The monument commemorating the thwarted passage of Napoleon III was probably moved during construction (Figure 4.1), and it is not even mentioned in the municipal archives. But the monument, although discreet, is still on the ground, in front of the town hall, a reminder of the French presence (Figure 4.1). It is next to the monument to the dead, another classic symbol of belonging to France.

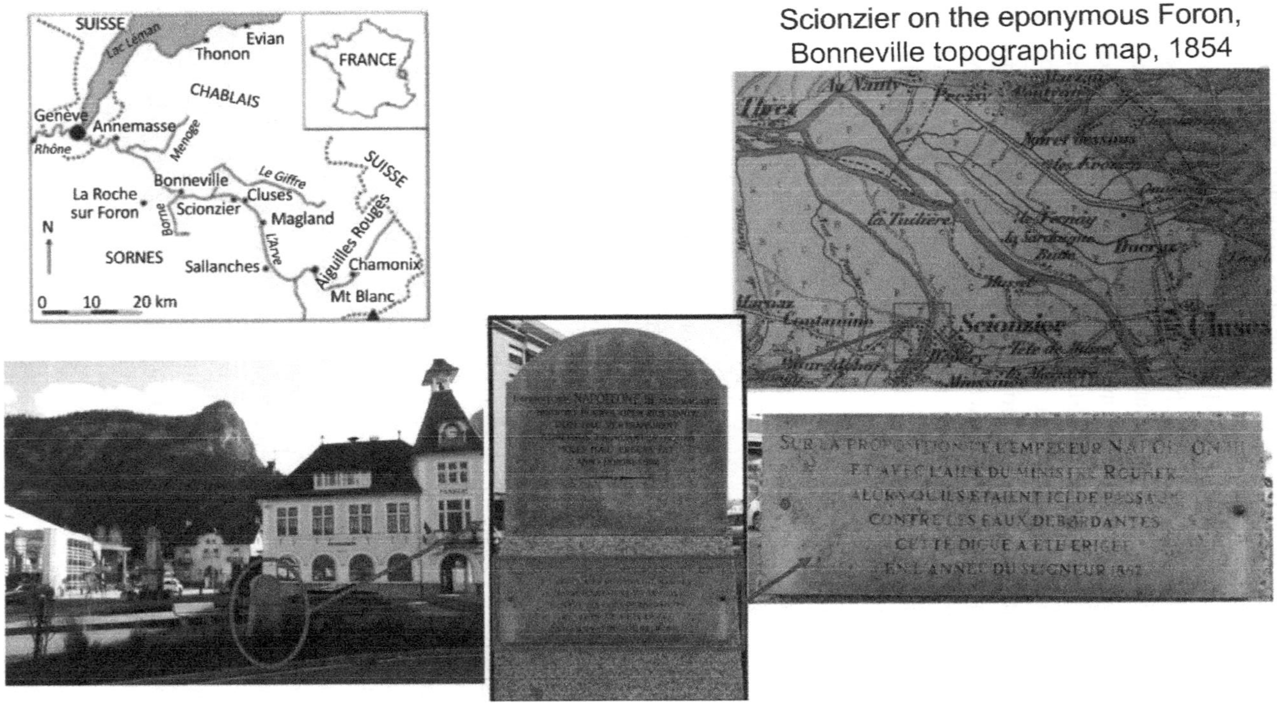

Figure 4.1. *RMT policy display in front of the town hall of Scionzier (Haute-Savoie) (source: P. Pigeon, February 2020. Topographic map of Bonneville, reference 1 FI 417, Archives départementales de la Haute-Savoie). For a color version of this figure, see www.iste.co.uk/rebotier/risks.zip*

The RMT policy decided from Paris was also a tool of political control, linked to the annexation of Savoy by France in 1860. The stele bears the following epigraph: "On the proposal of Emperor Napoleon III, and with the help of Minister Rouher, while they were here passing through, against the overflowing waters, this dam was erected, in the year of the Lord 1862."

The evolution of RMT policies toward the control of urbanization cannot be understood without the association between the State, insurance companies and mutual insurance societies that the Tazieff law (1982) put in place, and which has persisted since. It is then a question of displaying the risk using its official mapping, carried out by the State, to avoid the re-examination of responsibility (as specified in the decision of the Council of State, C.E. July 27, 1979 Sieur Blanc[1], Légifrance 2019[2]). It is also a question of creating a system that allows the insurance of so-called natural risks by mutualizing the premiums collected on a national scale, a system designed and decided from the capital (Pigeon 1991; Gérin 2010). The changes in RMT policies since then, and even more so those linked to the Tazieff law, PERNP (plans for exposure to predictable risks) and subsequent risk prevention plans, also revolve around the desire to maintain an economically viable, and therefore profitable, system.

4.3.3. *From one century to the next, managing risks to justify the power structure?*

We deduce that metropolitan policies, whether those of RMT or others, whether in reference to the Anthropocene or not, pursue other objectives than the sole prevention, however official and displayed, of risks associated with a type of hazard. This is nothing new. The insert taken from Braudel's reflections (1979, 1986) argues that the supply of the London metropolis alone has given rise to national policies in conjunction with traders, shipowners and insurers, while passing through the *private market*. The risks linked to supply disruptions are also geopolitical and, equally, they are financial risks.

---

1 "The Prefect having abstained until 1970 from implementing the procedure of delimitating zones exposed to natural risks, provided for by article 2 of the decree of August 29, 1955, in a high, mountainous sector promised to a significant urbanization and of which it does not appear from the documents of the file that it could be considered at the time as exempt from avalanche risks. This abnormal delay constitutes a fault which did not put the administration in a position to appreciate the reality and the importance of the risk when the building permit for a hotel-restaurant was issued in 1969 in this sector."

2 www.legifrance.gouv.fr/affichJuriAdmin.do?idTexte=CETATEXT000007667634 [Accessed 30 April 2020].

> These cities [the capitals] represent enormous expenses, their economy is balanced only from the outside, others must pay for their luxury. So, what are they good for in this West where they emerge and impose themselves so powerfully? They make the modern States, an enormous task, an enormous burden. They mark a turning point in the history of the world. They create these national markets, without which the modern state would be a pure fiction. For in truth, the British market was not born solely because of the political union between England and Scotland (1707), because of the *act of union* with Ireland (1801), or because of the beneficial abolition of so many tolls, or of the animation of transport, or of the "madness of the canals" or of the sea, which promotes free-trade by nature and envelops the islands, but rather of the ebb and flow of goods to and from London, the enormous demanding heart that gives rhythm to everything, moves and calms everything. Add the cultural, intellectual and even revolutionary role of these hothouses: it is enormous. But it is paid for, it is paid for at a very high price. (Braudel 1979, p. 465)

> This *private market*, which, well before the 18th century, began to organize throughout England direct purchases, often in advance, from producers, the purchase of wool, wheat, cloth, etc., from peasants outside the market, was the setting up, against the traditional regulation of the rules of the market, of autonomous commercial chains, very long, free in their movements, and which, moreover, took advantage of this freedom without scruple. They have imposed themselves by their efficiency, thanks to the large supplies needed by the army or the big capitals. The "belly" of London and the "belly" of Paris were in short revolutionary. In short, the 18th century developed everything in Europe, including the "counter-market". (Braudel 1986, p. 31)

It can, therefore, be argued that metropolises, through the multiplication of needs and issues linked to the increased concentration of populations and activities in a reduced space, multiply risk situations, but also facilitate their identification as well as the implementation of policies that claim to manage them. These policies dictate the taking into account of supralocal logics: they contribute to the transformation of planet Earth, without being able to control all the components of these transformations. Once again, the idea of the world being closed off behind the Anthropocene is not new, and is intrinsically linked to metropolitan settlements.

Expanding the perspectives further, Scott (2017) sees the politics of historical city-states as powerful means of control, of self-justification, aimed at reducing opposition, while relying on institutionally associated sets of actors. This capacity to control through norms presented as self-evident is one of the characteristics of ideologies. Ricoeur (1986, p. 32) argues that "it is the role of ideology to legitimize authority. More precisely, while ideology serves... as a code of interpretation that ensures integration, it does so by justifying the present system of authority". Ideology is at work both in academic work and in policy developments on disaster risk reduction. This tendency is not however new. It necessarily raises the question of the ideological role in the use and dissemination of the Anthropocene notion as a framework of control.

4.4. The Anthropocene, an ideological amplifier of responsibility transfers?

There is a basic tendency in research programs, theses and HDRs (*habilitations à diriger la recherche*[3]) to justify increased recourse to local actors when it comes to preventing disaster risks. Theses and HDRs on risk prevention, such as those of the author of this chapter (Pigeon 1991, 2002), did not insist on this aspect. However, one of the basic definitions of risk, that of CRED, already drew attention to the overcoming of local management capacities. Such a criterion suggests the benefit of strengthening local capacities for prevention purposes.

This observation of a recent interest in local capacity building is consistent with the syntheses on the ambivalent notion of *empowerment*: they also point to the very late transfer of this notion to French academic thinking, well beyond the field of research on disaster prevention or the Anthropocene. Bacqué and Biewener (2013) go so far as to mention the "hexagonal import" of *empowerment*, the notion arriving in France "in the 2000s". This trend is supported by international institutions, primarily by the UNDRR. It is ideological because it is only partly defensible on the logical level. Indeed, it is just as much conditioned by choices that remain debatable. This reading seems to us to be strongly consistent with that proposed by Althusser, quoted by Ricoeur (1986, p. 186): "An ideology is a system (possessing its own logic and rigor) of representations (images, myths, ideas or concepts as the case may be) endowed with an existence and a historical role within a given society".

3 Research thesis constituting a summary of completed work or an original proposal, and opening, in the French higher education and research system, to the possibility of supervising doctoral students.

4.4.1. *Locally focused research programs and risk management policies*

The tendency to value the capacity of local actors to prevent disaster risks accompanies the Anthropocene, in the sense that it emphasizes human responsibilities and transfers them, often in an undifferentiated way, to the local scale and to individuals. It is found at work in very different research programs, such as the European FP7 programs KNOW-4-DRR (Knowledge For Disaster Risk Reduction) or ESPON BRIDGES, to which the author of this chapter has contributed. The first concerns the relationship between knowledge and structures of research programs, on the one hand, and prevention policies associated with disaster risks, on the other hand. The second concerns the critical analysis of European aid policies for regions in structural difficulty. It includes the contribution of development to disaster risk prevention.

In both cases, there are differences in approach and apparently in theories. However, recommendations from academic reflections, including managers, also emphasize the need to value local knowledge more and to involve local actors more in policy-making (Weichselgartner and Pigeon 2015; Weichselgartner *et al.* 2016). These recommendations are partly defensible on their merits, as we saw above with the analysis of CRED's proposed definition of a disaster (section 4.1.2). They may also appear to be ethically defensible, as challenging the benefit of strengthening local decision-making, *a priori*, seems liberticidal.

There is no shortage of theses and HDR dissertations that pick up on this trend, linked to international policy frameworks, such as, first and foremost, the Sendai framework (UNISDR 2015)[4]. Research on topics as apparently different as the prevention of technological disasters (Propeck-Zimmermann 2015) or those related to marine submersions (Coquet 2019) draw attention to the benefit of strengthening prevention policies by giving more space to the knowledge of local actors. Such an approach would make it possible to reduce the discrepancies between what is expected of existing prevention policies, presented as top-down, and the many limitations they encounter on the ground. Contextualization (Michellier *et al.* 2016) and empowerment are thus presented as fundamental levers for strengthening disaster prevention. Moreover, other recent HDR dissertations point in the same direction, regardless of the types of disaster risks considered (Amalric 2019; Provitolo 2019).

Basically, it may seem logical to draw attention to the development of local resources when it comes to understanding and preventing disaster risks. According

4 Principles notably recalled at the UNISDR global platform held in 2019 in Switzerland: www.unisdr.org/conference/2019/globalplatform/home [Accessed 16 June 2020].

to the classic definition of CRED (2019), a disaster is an event that exceeds local management capacities. Strengthening these capacities, therefore, seems an inescapable necessity. At the same time, however, this logical necessity seems to call into question the top-down disaster risk prevention policies associated with the transfer of resources from actors operating on a supralocal scale, namely the States and European institutions, in the first instance. Recourse to states would then remain justified in the form of emergency aid, as is the case, for example, in Brazil even today (De Oliveira Magalhaes 2019).

4.4.2. *The importance of fully pursuing the critique so as to avoid becoming trapped by ideology*

But such a position, which may seem logical, quickly finds its limits, and thus becomes ideological when it continues to be asserted indiscriminately. What is the sense of building local capacity in the case of low-density settlements, or those associated with households of limited socioeconomic means? This is one of the recommendations of the ESPON-BRIDGES report. Even if the resources are not necessarily or only monetary (Sen 2010), one can quickly see the limits of the argument: what can *empowerment* mean with so few local resources available? Bacqué and Biewener (2013) begin to provide an answer. They distinguish three models of *empowerment*: radical, liberal and neoliberal. In the latter case, "the notion of *empowerment* is mobilized within a logic of poverty and inequality management, to enable individuals to exercise their own capacities and make 'rational' decisions in a market economy context... without calling into question social inequalities".

Moreover, giving more decision-making power to local actors in order to strengthen disaster risk prevention is also a solution that has its limits. The refusal of risk prevention plans, for a commune like La Faute-sur-Mer, contributed to the preparation of the February 2010 disaster (Pigeon 2012b). It can be explained above all by the municipal political choices made by the majority of actors, elected officials and local voters. They opposed a territorial planning tool, the risk prevention plan, prescribed by the prefect, the representative of the State in the department. Through the red zonings, which cannot be built on, the risk prevention plans reduce the number of buildings exposed. Through blue zonings and architectural prescriptions, the risk prevention plans aim to reduce the vulnerability of these buildings as well as the associated populations and activities. In this case, the state tool aims to prevent disasters, while the municipality, relying on information from local citizens, takes decisions that go in the opposite direction. This was verified by the disaster of February 2010: the date was marked by the death by drowning of 29 people in a single night, during the passage of the storm Xynthia. These victims were trapped in their houses, built without an upper floor and in a flood zone, in accordance with the regulations of the local urban plan. Could the helicopter that came to save some of the inhabitants who were able to find refuge on the roof of their house have been financed by a budget allocated to a commune of a little more than a thousand inhabitants in 2010?

Finally, it is surprising to see such solutions presented as innovative. The empowerment of the individual is in no way a contemporary discovery of the Anthropocene moment or a response to new environmental emergencies. As Larrère and Larrère (1997, p. 207) point out, the interpretation of the Lisbon disaster in 1755 facilitates identification of human responsibility, without this necessarily leading to moral condemnation. Even more, Rousseau was not so innovative here: "In this reflection, Rousseau implements a scheme that he borrows from Malebranche, and which leads him to impute to men the misfortunes they complain about". With this controversy, we are at the formal beginnings of liberalism, in the middle of the Enlightenment and on the eve of the success of thinkers like Benjamin Constant. The Lisbon controversy on the imputability of responsibilities is well on the threshold of the advent of liberal ideas... and then neo-liberal ideas much later.

As for the neo-liberal currents that more recently question the transfer of policies from the State, they are no more radically new. One of their roots is in the United States in connection with the Tea Party: the expression refers to the contestating of decisions coming from London (the pre-revolutionary uprising in Boston), but also to the contestating of taxes associated with public authorities, led by the federal state. The valorization of private enterprise and the individual was even more profoundly carried by certain religious currents, notably those linked to Calvinism. And these, as Braudel (1986) and especially Weber (1905/2017) remind us, have elective affinities with the different forms of capitalism. These different elements were recomposed and taken up again, supporting the intensification of globalization, during the 1980s, under the main influence of the governments led by Margaret Thatcher and Ronald Reagan. Many readings of the Anthropocene carry within them these multiple conditionings, some distant, others more recent, while recomposing them.

4.5. Back to the future: the Anthropocene as a new avatar of "creative destruction"?

On the points set out in this chapter, it can therefore be argued that there is nothing fundamentally new about the Anthropocene. It is mainly a matter of intensifying the recognition of risks created essentially by human societies, and recognized by metropolitan actors, in order to justify and recompose pre-existing policies, and in particular policies for the prevention of disaster risks. It is not only a question of injunctions to manage these risks of disaster, or even catastrophe, by relying on the perspectives opened up by *collapse studies*. It is also a question of reconsidering the prior norms, which are also instruments of political control, accompanying the evolution of markets. In essence, we find the analysis of Braudel and the London market, with the *private market*, and even Schumpeter's analyses in terms of creative destruction (1947).

However, to what extent does the intensification of the mutations of the planet and of human populations not also announce the limits of this type of evolution, and of more fundamental transformations? Here again, the debate can no more be settled than when it was posed by Boserup and Condorcet, or by Malthus... The major difference between the two thoughts is that Malthus assumes that demographic evolution is not very closely linked to the evolution of resources, whereas Boserup "showed how demographic growth (pressure) led to a reorganization of the collection and production of food" (Le Bras 1994, p. 254).

Le Bras is more on the side of Boserup than Malthus, insisting, for example, on the political character of famines. But, at the same time, as the title of his book announces, the world is also finite: certainly not as finite as the world of flies placed in a jar during a biology experiment, but finite nonetheless, adaptive capacities or not. The coupling of demographic and resource developments is therefore also finite, as Malthus assumes. But in what terms are the couplings, the interdependencies, the solidarities and their unequal distribution in space and among societies posed? These are many of the issues at stake in a questioning of risks and the environment rooted in politics.

It can therefore be argued that the problems posed in the Anthropocene moment are nothing more than a recomposition of old thoughts: a resumption of a very old debate on the relations between humanity and planet Earth, which does not cut across and reconsider the dominant political and ideological structures.

However, the success of the use of the Anthropocene can equally mark recognition of the limits posed by the extension and intensification of a form of globalization, heralding catastrophic transformations for humanity in a world that is factually closed.

4.6. References

Aitsi-Selmi, A.V., Murray, V.C., Wannous, C., Dickinson, C., Johnston, D., Kawasaki, A., Stevance, T., Yeung, T. (2016). Reflections on a science and technology agenda for 21st century disaster risk reduction. *Int. J. Disaster Risk Sci*, 7(1), 1–29 [Online]. Available at: www.preventionweb.net/publications/view/49515 [Accessed 30 April 2020].

Amalric, M. (2019). Habiter l'environnement, pour une géographie sociale environnementale : nature, paysage, risque. HDR, Université de Bretagne Occidentale, Brest.

Arnould, B., Demolis, C., Douard, P., Meignien, X. (2016). La restauration des terrains en montagne (RTM). Mise en œuvre de la politique de prévention des risques par les services RTM. Report, Conseil général de l'environnement et du développe-ment durable, Paris [Online]. Available at: www.actu-environnement.com/media/pdf/news-26543-rapport-CGEDD-restauration-terrains-montagne.pdf [Accessed 30 April 2020].

Bacqué, M.H. and Biewener, C. (2013). L'empowerment, un nouveau vocabulaire pour parler de participation? *Idées économiques et sociales*, 173(3), 25–32 [Online]. Available at: www.cairn.info/revue-idees-economiques-et-sociales-2013-3-page-25.htm# [Accessed 30 April 2020].

Beck, U. (1992). *Risk Society: Towards a New Modernity*. Sage Publishing, London and New York.

Bergson, H. (2014). *La pensée et le mouvant*. Flammarion, Paris.

Bonnet, J. (1994). *Les grandes métropoles mondiales*. Nathan Université, Paris.

Braudel, F. (1979). *Civilisation matérielle, économie et capitalisme*, Volume 1. Armand Colin, Paris.

Braudel, F. (1986). *La dynamique du capitalisme*. Flammarion, Paris.

Bravard, J.P. (1987). *Le Rhône*. La Manufacture, Lyon.

Brugnot, G. and Cassayre, Y. (2003). De la politique française de restauration des terrains en montagne à la prévention des risques naturels [Online]. Available at: www.fao.org/3/XII/0034-C3.htm#P84_20925 [Accessed 30 April 2020].

Carlier, B. (2019). Les risques d'origine torrentielle et gravitaire dans la haute vallée du Guil (Queyras, Alpes du Sud, France). PhD Thesis, Université de Paris 7, Paris.

Cerema (2019). Sécurité globale et résilience des territoires – Ingénierie effets de leviers et stratégies à promouvoir. Centre d'études et d'expertise sur les risques, l'environnement, la mobilité et l'aménagement, Bron.

Combe, C. (2007). La ville endormie ? Le risque d'inondation à Lyon : approche géohistorique et systémique du risque de crue en milieu urbain et périurbain. PhD Thesis, Université Lumière, Lyon.

Comby, E. (2015). Pour qui l'eau ? Les contrastes spatio-temporels des discours sur le Rhône (France) et le Sacramento (États-Unis). PhD Thesis, Université de Lyon 3, Lyon.

Coquet, M. (2019). La perception du risque de submersion marine par les habitants de communes littorales françaises : les cas de Barneville-Carteret, Saintes-Maries-de-la-Mer, Châtelaillon-Plage et Sainte-Anne (Guadeloupe). PhD Thesis, Université de Nantes, Nantes.

CRED (2019). Base de données EM-DAT, emergency database. Centre de recherches sur l'épidémiologie des désastres, Louvain-la-Neuve [Online]. Available at: www.emdat.be/ [Accessed 30 April 2020].

David, M. (2016). Dynamique fluviale de la Garonne à l'anthropocène : trajectoire d'évolution du tronçon fluvial compris entre les confluences de l'Ariège et du Tarn (Garonne toulousaine, 90 km). PhD Thesis, Université de Toulouse 2, Toulouse.

DCE (2000). Directive n° 2000/60/CE du 23/10/00 établissant un cadre pour une politique communautaire dans le domaine de l'eau [Online]. Available at: https://aida.ineris.fr/consultation_document/995 [Accessed 16 June 2020].

De Oliveira Magalhaes, A. (2019). Aléas et risques naturels (inondations, mouvements de terrain) dans le Nordeste du Brésil. Une approche géographique appliquée à la Région du Cariri (Crato et Barbalha, Ceará). PhD Thesis, Université de Paris 7, Paris.

Diamond, J. (2005). *Effondrement. Comment les sociétés décident de leur disparition ou de leur survie.* Gallimard, Paris.

Drouin, J.M. (1993). *L'écologie et son histoire.* Flammarion, Paris.

Farinelli, B. (2015). *La révolution de la proximité. Voyage au pays de l'utopie locale.* Libre et solidaire, Paris.

Fressoz, J.B. and Bonneuil, C. (2013). *L'événement anthropocène, la terre, l'histoire et nous.* Le Seuil, Paris.

Gérin, S. (2010). Une démarche évaluative des Plans de Prévention des Risques dans le contexte de l'assurance des catastrophes naturelles : contribution au changement de l'action publique de prévention. PhD Thesis, Université de Paris 7, Paris.

Jonas, H. (1979). *Le principe responsabilité.* Le Seuil, Paris.

Larrère, C. and Larrère, R. (1997). *Du bon usage de la nature : pour une philosophie de l'environnement.* Flammarion, Paris.

Le Bras, H. (1994). *Les limites de la planète.* Flammarion, Paris.

Magny, M. (2019). *Aux racines de l'anthropocène, une crise écologique reflet d'une crise de l'homme.* Bord de l'eau, Paris.

Menoni, S. and Margottini, C. (2011). *Inside Risk: A Strategy for Sustainable Risk Mitigation.* Springer, Berlin.

Michellier, C., Pigeon, P., Kervyn, F, Wolff, E. (2016). Contextualizing vulnerability assessment: A support to geo-risk management in central Africa. *Natural Hazards*, 82(1), 27–42.

Morin, E. (1977). *La méthode, tome 1 : la nature de la nature.* Le Seuil, Paris.

Morin, E. (2014). *Introduction à la pensée complexe*. Le Seuil, Paris.

Mougin, P. (1914). *Les torrents de Savoie*. La Fontaine de Siloé, Chambéry.

Pigeon, P. (1991). L'homme face aux risques dits naturels dans le massif du Chablais et son piedmont (1860–1990). PhD Thesis, Université de Grenoble 1, Grenoble.

Pigeon, P. (1994). *Ville et environnement*. Nathan Université, Paris.

Pigeon, P. (2002). Réflexions sur la géographie des risques dits naturels. Habilitation à diriger des recherches, Université de Nice-Sophia-Antipolis, Nice.

Pigeon, P. (2012a). *Paradoxes de l'urbanisation*. L'Harmattan, Paris.

Pigeon, P. (2012b). Apports de la résilience à la géographie des risques : l'exemple de La Faute-sur-Mer (Vendée, France). *[VertigO] La revue électronique en sciences de l'environnement*, 12(1) [Online]. Available at: https://doi.org/10.4000/vertigo.12031 [Accessed 30 April 2020].

Pigeon, P. and Rebotier, J. (2016). *Disaster Prevention Policies: A Challenging and Critical Outlook*. ISTE Press, London, and Elsevier, Oxford.

Pigeon, P., Rebotier, J., Guezo, B. (2018). Ce que peut apporter la résilience à la prévention des désastres : exemples en Lavours et en Chautagne (Ain, Savoie). *Annales de géographie*, 719, 5–28.

Propeck-Zimmermann, E. (2015). Modélisation des risques et décision territoriale. Recherche sur les risques socio-techniques en milieu urbain. HDR, Université de Caen, Caen.

Provitolo, D. (2002). Risque urbain, catastrophes et villes méditerranéennes. PhD Thesis, Université de Nice-Sophia-Antipolis, Nice.

Provitolo, D. (2019). Des modèles multiples pour l'étude des risques et des catastrophes. HDR, Université Côte d'Azur, Nice.

Reghezza, M. (2007). Réflexions autour de la vulnérabilité métropolitaine : la métropole parisienne face au risque de crue centennale. PhD Thesis, Université de Nanterre-Paris 10, Paris.

Revault d'Allonnes, M. (1999). *Le dépérissement de la politique*. Flammarion, Paris.

Ricœur, P. (1997). *L'idéologie et l'utopie*. Le Seuil, Paris.

Schumpeter, J. (1990). *Capitalisme, socialisme et démocratie*. Payot, Paris.

Scott, J.C. (2017). *Homo domesticus. Une histoire profonde des premiers États*. La Découverte, Paris.

Sen, A. (2010). *L'idée de justice*. Flammarion, Paris.

Simon, D. (2016). Quand un concept écologique fait date. L'invention du "complexe pathogène" en géographie. *Revue d'histoire des sciences humaines*, 28, 253–272.

UNISDR (2015). Global assessment report on disaster risk reduction [Online]. Available at: www.undrr.org/publication/global-assessment-report-disaster-risk-reduction-2015 [Accessed 16 June 2020].

WB (2019). The Anthropocene: From global change to planetary stewardship. *World Bank*, Washington [Online]. Available at: https://datacatalog.worldbank.org/anthropocene-global-change-planetary-stewardship [Accessed 30 April 2020].

Weber, M. (2017). *L'éthique protestante et l'esprit du capitalisme.* Flammarion, Paris.

WEF (2017). Welcome to the Anthropocene. *World Economic Forum*, Geneva [Online]. Available at: www.facebook.com/worldeconomicforum/posts/welcome-to-the-anthropocene/10154256938321479/ [Accessed 30 April 2020].

Weichselgartner, J. and Pigeon, P. (2015). The role of knowledge in disaster risk reduction. *Int. J. Disaster Risk Sci.*, 6, 107–116 [Online]. Available at: https://doi.org/10.1007/s13753-015-0052-7 [Accessed 30 April 2020].

Weichselgartner, J., Norton, J., Chantry, G., Brévière, E., Pigeon, P., Guézo, B. (2016). Culture, connaissance et réduction des risques de catastrophe : liens critiques pour une transformation sociétale durable. *[VertigO] La revue électronique en sciences de l'environnement*, 16(3) [Online]. Available at: www.erudit.org/en/journals/vertigo/2016-v16-n3-vertigo03075/1039993ar.pdf [Accessed 30 April 2020].

WMO (2017). The Anthropocene. *World Meteorological Organization*, Geneva [Online]. Available at: https://theanthropocene.org/press/2017/record-breaking-climate-change-pushes-world-into-uncharted-territory/ [Accessed 30 April 2020].

5

The International World of Disasters: Beyond Reflexivity, Surpassing Naturalism?

Sandrine REVET
CERI, Sciences Po – CNRS, Paris, France

5.1. Introduction

In 2019, the UN agency in charge of disaster risk reduction (UNDRR)[1] produced a "Global Review" whose objective was to highlight the progress made in implementing the Sendai Framework for Action (2015–2030). This 472-page report does not mention the notion of the "Anthropocene" at any point. On the same UNDRR website, the "terminology" section does not have an entry for "Anthropocene" either. However, this section includes the glossary recently revised by professionals in the international world of "natural" disasters, and in particular by all those involved in risk reduction or the management of so-called "natural" disasters at the international level (Revet 2018)[2]. How can we understand this absence,

1 https://gar.undrr.org/.

2 This chapter is based on a multi-site ethnographic survey conducted from 2008 to 2015 among professionals in the international "natural" disaster world. It is based on ethnographic observations, a series of 43 interviews and the analysis of numerous documents and events organized by professionals in this world. This survey, financed mainly by the CERI laboratory and by the Scientific Direction of Sciences Po, led to the publication of a book (Revet 2018), recently translated into English (Revet 2020). Some passages of this text have already been published in this book.

Risks and the Anthropocene,
coordinated by Julien REBOTIER. © ISTE Ltd 2021.

when the Anthropocene has become an inescapable notion when we talk about global warming[3] or the world environmental crisis? In order to shed light on this question, we can undoubtedly try to return to the types of knowledge that have been at the heart of the production of the international world of disasters and to the types of relations between humans and nature on which this knowledge is based. Because, in the end, it is the terms of this relationship that are at stake, notably through the way in which responsibilities are attributed in the crises that emerge between humans and nature. Explaining "natural" disasters by the magnitude and force of a telluric or climatic phenomenon, or by the social conditions of vulnerability of the affected territory and society, means engaging different systems of attributing responsibility and thereby adopting a "framing" (Revet 2018) that leads to linking a way of understanding what is happening with a way of acting. These framing operations are part of a certain type of relationship with nature.

In this chapter, I will defend the idea that since its emergence in the 1970s and until very recently, the international world of "natural" disasters has been part of a "naturalist" (Descola 2005) understanding of the world, which asserts a separation between "nature" and "culture" and for which society is an exclusively human affair (Descola 2018). Whether it is Earth scientists who have been at the heart of the process of internationalizing this world since the 1970s, or the social scientists who have imposed their analyses since the 1990s, the scientists who have contributed to producing the various framings of this international world have only very recently begun to think beyond this naturalism to consider other kinds of relationships between nature and humans, which would be based on interdependence and interrelation rather than on a radical separation.

Another factor explaining the relative absence of the notion of the Anthropocene in the world of international disaster is the very nature of the phenomena that this world faces. In most cases, "natural" disasters remain relatively localized, unlike other types of disasters, such as pandemics or technological disasters of the nuclear type, whose effects spread beyond national borders and whose international dimension is clearly apparent. It is with this point that I propose to begin this chapter, before turning to the different types of knowledge that have accompanied the production of an international world of "natural" disasters.

---

3 I prefer to speak of global warming rather than climate change in order to emphasize the anthropogenic dimension of greenhouse gas emissions and the global, planetary dimension (see Morton 2013). However, in the international world I study, actors tend to use the term "climate change", which I will therefore use again in this chapter, depending on the context.

5.2. Localized disasters dealt with internationally

Unlike pandemics or global warming, whose global character is ontological, so-called "natural" disasters remain localized events. Their effects rarely extend beyond the borders of a country, and even if their causes are sometimes linked to broader global phenomena such as global warming or development problems, the link is often not clearly visible.

Moreover, states remain sovereign in their decision to request international assistance after a disaster and are the central actors in emergency response. The international character of disasters, and even their transnational character when it comes to going beyond the level of state action, is therefore not obvious. It is the result of a construction work that involves at the same time discourses, practices and devices that link together humans, tools and words. This internationalization has taken institutional forms, with the creation of UN agencies specialized first in the coordination of disaster relief actions set up by UN actors (UNDRO[4] in 1971, DHA[5] in 1991 and then OCHA[6] in 1998), then with the establishment of an inter-agency system specialized in disaster prevention (International Decade for Natural Disaster Reduction 1990–1999), and then in risk reduction (UNISDR[7], at present UNDRR[8]). This process has been supported by the production of discourses, good practice guides, standardization tools, courses, training and databases that have made it possible to turn these territorialized phenomena into "international" disasters, that is, handled by a set of international and transnational actors who speak the same language, act in a coordinated manner and use the same protocols and devices.

However, the establishment of this international system of management and prevention has not gone hand in hand with a reconsideration of the interpretation that consists of analyzing "natural" disasters from their place of occurrence, that is, a local, urban or rural, more or less vast space. The local manifestations of these phenomena have taken precedence over any search for global causes. Although since the 2000s it has been common to associate the causes of climatic phenomena with the change of the same name and consequently to link disaster risk reduction and adaptation to climate change (Kelman *et al.* 2017), this movement is relatively recent.

---

4 United Nations Disaster Relief Organization.

5 United Nations Department of Humanitarian Affairs.

6 United Nations Office for the Coordination of Humanitarian Affairs.

7 United Nations International Strategy for Disaster Risk Reduction.

8 United Nations Office for Disaster Risk Reduction.

This can be seen, for example, in the images used in international disaster reporting to represent disasters. In addition to images representing victims and the most vulnerable (women, children), the covers of reports and the websites of agencies specialized in disaster management and risk reduction are often adorned with photos representing hazards. These hazards are visible, tangible phenomena, often large in size, and are therefore easily represented, unlike other less visible, or even totally invisible hazards, which concern other types of disaster (a radioactive cloud, chemical pollution) that most of the time force one to choose to represent the source of the hazard (the factory, the nuclear plant, the reactor[9]). The images of a volcanic eruption, of a crack caused by an earthquake, of a 20-m wave during a tsunami, of a ground cracked by a drought, of a torrent of mud or of a lightning flash all contribute to produce a spectacular, esthetic and universal iconography of the "natural" catastrophes. In this iconography, the importance of hazards is striking. Indeed, representations of the phenomenon itself are most often associated with those that show the material destruction it causes: boats displaced inland (Figure 5.1), cars covered in water, collapsed buildings, a portion of road deformed by an earthquake, etc. The disaster appears then in the disorder that it creates, because it mixes what should be separated; it soils, breaks, destroys (Langumier 2008). This iconography can be qualified as naturalist (Descola 2005) in that it represents a nature radically separated from the humans that it affects. It can also be qualified as animist by certain aspects insofar as, insisting on its destructive character, it also displays a disturbing nature, provided with *agency* and interiority: of a capacity to take action on its own account and a conscience. Nature is staged as the culprit for the deaths and damages it causes.

Other images respond in some way to this "hazard-centered" iconography – that is, one that places the natural phenomenon at the center. It is a production that can be qualified as modernist, which represents the means by which man controls the planet. This iconography is nourished by images of planet Earth, seen from the sky, in the form of a globe or a planisphere, photos of satellites or objects moving in space. This perspective is also representative of a naturalist reading of the relationship between humans and nature. It refers directly to imagery of the Cold War, the conquest of space and the ideology that humans can and should control natural phenomena (Hamblin 2013). The chosen scale, in contrast to the first type

9 As we can see with the COVID-19 pandemic, viruses, although small and invisible to the naked eye, do not escape representation. An iconographic analysis of the production of articles and reports on this phenomenon would undoubtedly reveal the artistic dimension of the representations of this round virus with its crown of colored peaks.

of images, is striking in its immensity and excess. If with the preceding images, one was situated at the scale of the phenomenon itself, localized, territorialized, modernist iconography tackles issues on the scale of the globe, without being situated in a thought that exceeds naturalism. At this scale, humans appear only in the technological means – instruments of observation, measurement and anticipation – that they deploy to control nature and its disturbances (Figure 5.2).

Figure 5.1. *Boat perched on the roof of a house in Banda Aceh, Indonesia, after the tsunami of December 26, 2004 (source: Franck Lavigne/LGP/CNRS Photo Library)*

This modernist representation of the planet is also present in the iconography of global warming, in which the globe, the planet, is at the heart.

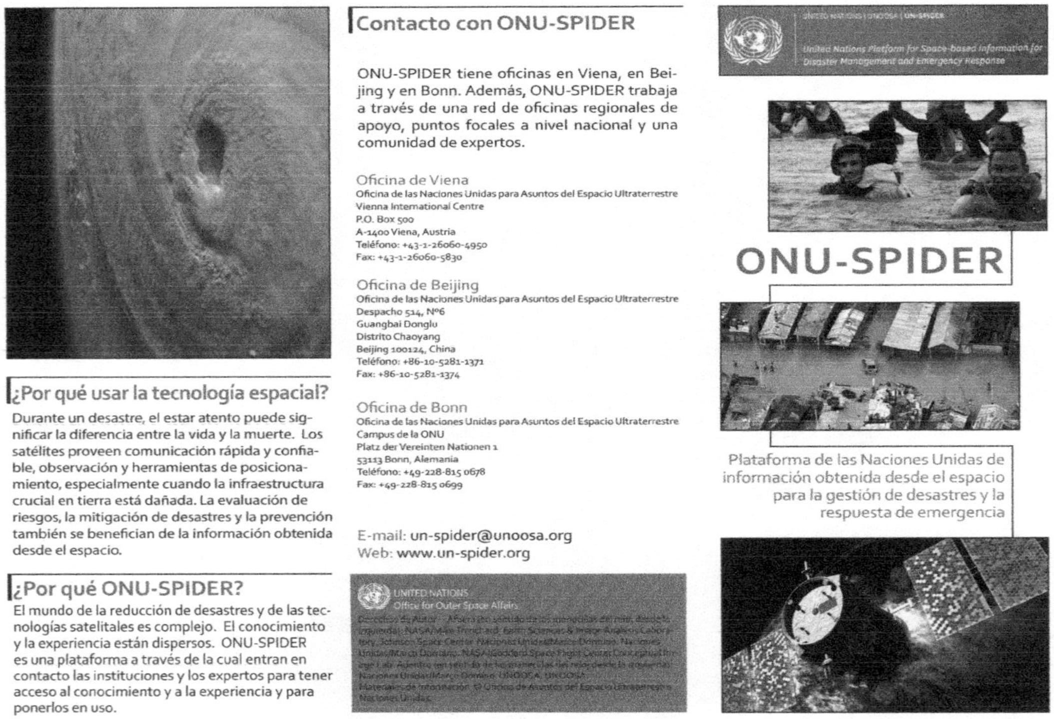

Figure 5.2. *Flyer presenting the UN-SPIDER organization in Spanish*

However, the theme of global warming sometimes also allows us to ask other questions, particularly in terms of responsibility and justice, and in a way to question the purely naturalist framing. We see, with global warming, an iconography emerging, not only focused on victims or hazards, but also on questions of rights, justice and responsibilities (Manzo 2010). Kate Manzo, for example, analyzes the 2007 campaign of the NGO Christian Aid, which depicts a Nigerian goat herder crouching on the ground, facing the camera, surrounded by his dead goats and calling out to the inhabitants of the richest countries the consequences of their excessive heating on the death of his herd (Manzo 2010). Here, the link is established between the actions of certain humans and the consequences of these actions on the environment and on the lives of other humans and animals, geographically, culturally and socially distant. The image narrows the distance between them and allows them to meet. In this iconography, it is therefore primarily through the question of justice and responsibility that the relationship between humans and nature is reinterrogated. The accent is put on the relationship, on the interdependencies and not on the radical separation as it may be in the naturalistic perspective. This naturalist perspective is, as we shall see, predominant among professionals in the world of "natural" disasters.

5.3. Disaster science: a naturalistic framework

In order to understand the importance of this naturalist framework in the international world of "natural" disasters, it is important to look back at the history of the sciences that have accompanied and participated in its production and in the internationalization of disasters. For a long time, "natural" disasters were thought of only from the point of view of the natural sciences, which focused on the phenomena themselves. Earth sciences and climate sciences then shared the stage in an attempt to understand natural hazards such as earthquakes, volcanic eruptions, hurricanes or storms. Lukas Schemper (2016) shows that the beginning of the 20th century was the occasion, at the same time as the first attempts to internationalize the theme of "natural" disasters and the creation of the International Relief Union, to set up the first international research and publication networks on disasters, their causes and their consequences. In particular, the creation of the journal *Matériaux pour l'étude des calamités* in 1924 with the help of the *Société géographique de Genève* is, according to him, an important but forgotten stage in the history of disaster studies. The works of geographers, mostly centered on the phenomena themselves, rubbed shoulders with the numerous works of seismologists, underlining the weight that this discipline would have in the development and study of "natural" disasters.

The period of the Cold War was then formidably favorable to the development of the Earth sciences, and the knowledge that was then developed was without precedent in terms of data and scale (Doel 2003). In order to understand how the different types of knowledge centered on the study of the Earth and climate – meteorology, oceanography, geophysics, terrestrial magnetism, hydrology, seismology, tectonics – build up knowledge about "natural" disasters, it is necessary to place them in the context of their extraordinary development during the 20th century. As Ronald Doel points out, the unification of multiple fields under the common name of "Earth sciences" during the 20th century was:

> one of the most important intellectual and institutional transformations in 20th century science, comparable to the unification of physiology, bacteriology and medical chemistry into the field of "biology" at the end of the 19th century. (Doel 2003, p. 391)

It is around the theory of plate tectonics, accepted at the end of the 1960s after years of scientific debate, and thus the rejection of the idea of the fixed nature of the Earth's structure, that a series of major transformations took place. The re-understanding of this phenomenon has allowed immense progress in our understanding of the formation of mountains or seismic zones, and has also contributed to making "natural" disasters relevant objects of study for scientists, insofar as their importance in the history of the Earth's evolution was considered afresh. Furthermore, the theory of plate tectonics is characterized by the fact that it allows a predictive perspective.

But it is also by resituating the extraordinary development of the Earth sciences in the more precise context of the Cold War that we understand both the factors that contributed to these transformations, and the impact that these developments may have had on the way of thinking about disasters and on the measures encouraged to deal with them. During the Cold War period, the interest of the military and governments, particularly in the United States, in the possible contributions of the Earth sciences was growing. It is around the negotiations related to the signing of the 1963 Nuclear Non-Proliferation Treaty that unprecedented possibilities for the Earth sciences emerge. It was also during the Cold War that a vast operation of general Earth and environmental monitoring was put in place (Hamblin 2013): to overcome the growing difficulties of spying on the Soviet enemy, Earth and climate sciences were put to work. Numerous monitoring networks – global radiological monitoring network, atmospheric and ocean monitoring, satellites – then made it possible not only to monitor the enemy but to produce a global image of the Earth through what Paul Edwards has designated as "global knowledge infrastructures" (Edwards 2010). Among these, a global earthquake monitoring network played a central role, the World Wide Standard Seismograph Network (WWSSN), set up at the end of the 1950s and during the 1960s with the help of UNESCO. The aim was to monitor seismic activity throughout the world in order to differentiate between natural earth movements and those caused by nuclear testing, with the aim of getting the two opposing camps to sign a non-proliferation treaty (Barth, 2001). The WWSSN was the first global data exchange system for earthquakes and earth movements (Cabane and Revet 2015). Seismologists were actively involved in this project, and in the United States, their discipline received unprecedented funding from the Department of Defense. In this way, seismology gained a prominent place, both in the scientific landscape and with political institutions, both in the United States and internationally (Barth 2003; Doel 2003).

5.4. Understanding nature in disasters

The framing that the Earth sciences and seismology in particular propose for thinking about "natural" disasters and for dealing with them is above all centered on understanding, measuring and even predicting the hazard – the natural phenomenon – itself. The means encouraged are consequently technological in nature and take little or no account of social factors, which some researchers have been considering for some years now as possibly contributing to worsening the effects of natural phenomena, as we shall see. This hazard-centered framing is not, however, the exclusive preserve of the Earth sciences (for the invasive nature of this framing, see Chapter 6), and the North American case offers us yet another example of the influence of certain research on the way of thinking and representing "natural" disasters. From the end of the 1940s onwards, social science research developed in the United States under the name of *Disasters Studies*. Funded by the U.S. Department of State and the National Academy of Sciences, the National Opinion Research Center (NORC) at the University of Chicago conducted the first systematic research on the behavior of populations in crisis situations between 1949 and 1954. Essentially centered on the North American field, this research was inspired by earlier works, notably those of Samuel Prince in 1920 on the explosion of a load of ammunition in Halifax harbor (Prince 1920), of Lowell Carr in 1932 on the impact of different components of culture on social responses to disasters (Carr 1932) and of Pitrim Sorokin in 1942 on the effects of different types of disasters on social organization (Sorokin 1942). The research conducted by the NORC Disaster Group aims to treat natural or technological disasters as "laboratory" situations to study what would happen in the event of a nuclear attack. This work focuses on the social consequences of catastrophic events, while considering that their causes – natural threats – are external to the society affected. They provide answers to the question of how people and societies perceive danger and react to disasters, and are in line with the hazard-centered perspective insofar as they focus on the social consequences of the phenomenon itself. The sociologist Charles Fritz, then director of this program, and researchers such as Enrico Quarantelli, Russell Dynes and J. Eugene Haas are interested, for example, in the eventualities of panic or in the factors that promote or prevent looting (Fritz and Marks 1954; Quarantelli 1954). From the 1960s onwards, this research was structured with the creation of the Disaster Research Center (DRC), first installed at Ohio State University and then at the University of Delaware and generously financed, from 1962 onwards, by the Office of Civil Defense.

Other works in the social sciences consider natural phenomena as "external" to society and insist on the "rare" and "violent" character of disasters, while at the same time, since the mid-1940s, they have been developing a reflection on the systems of adaptation or adjustment in the face of risks and disasters, from an "ecological" perspective. This work, grouped under the name of *Hazard Research*, was carried out in particular by the geographer Gilbert F. White with his thesis on flood management in the United States in the 1930s (White 1945) and then by his students such as Ian Burton and Robert Kates (Burton *et al.* 1968, 1978). These geographers innovated with respect to their predecessors insofar as they were interested in the different ways in which societies anticipate and respond to disasters. Their work, in a way, will inspire thinking around adaptation and resilience to global warming (Felli 2016).

5.5. Denaturalizing disasters or the arrival of reflexivity

In line with the importance of this research, which is primarily interested in the consequences of disasters or their physical causes, others have opened up new perspectives. Anglophone researchers in the social sciences, most of them Marxist-inspired geographers, such as Ben Wisner, Phil O'Keefe and Terry Cannon, laid the foundations of a reflection that defined itself as "radical", seeking to establish the link between disasters and modes of economic development (Wisner *et al.* 1977). These committed researchers took a stand against the dominant approaches – both in the Earth sciences and in American disaster studies – and proposed to analyze disasters from their causes, which were no longer natural, but historical, economic or political.

Their approach was marked by two major events that occurred in countries of the South: first, the great drought that affected Africa, particularly the Sahel zone (1967–1973), and led to deadly famines; second, the Bhola cyclone, which ravaged East Pakistan (now Bangladesh) in November 1970 and caused the death of several hundred thousand people. In this region, the management of the emergency by the government authorities was severely criticized and the civil war that followed led to the independence of Bangladesh. These events were widely publicized and led to the first major charity concert, organized by former Beatle George Harrison, on August 1, 1971 in New York.

In France, two works also take the opposite view of hazard-centered analyses: *Qui se nourrit de la famine en Afrique ?* in 1974, and *Sécheresse et famine du Sahel* in 1975. Refuting the idea that the two main causes of famine are the absence of rain and the irrationality of "traditional" social practices, the anthropologists Jean Copans and Claude Meillassoux want to think of drought as revealing a situation of economic and political domination. The famine is, according to them, the result of a colonial history and development policies that have led the populations of the region to a destructive dependence. Their perspective is indeed to re-politicize natural disasters and to "denaturalize" them. These works focus on an analysis of "vulnerability", of the "*root* causes" of disasters or of the "social construction of risk". For some, such as the Canadian geographer Kenneth Hewitt, it is a matter of showing that these phenomena described as natural are not events that break a "normal order" but, on the contrary, processes produced by an "abnormal" political, economic and social order (Hewitt 1983). While not leaving behind a naturalistic reading of these events insofar as nature and society are still thought of as radically separate entities, these analyses nevertheless introduce a first reflexive turn in the analysis of disasters, in that they situate their causes of disasters in anthropic activities and no longer in natural phenomena alone.

The 1990s saw the development of this vulnerability-based framework, with the publication of a collective work that was to play a central role in its international dissemination: *At Risk*, published in 1994, co-edited by Ben Wisner and his colleagues (Blaikie *et al.* 1994). This work marks the birth of what some people call the "vulnerability paradigm", which is being built against the "hazard-centric paradigm".

5.6. What framing at the international level?

In 1988, the United Nations finally took up the issue of "natural" disaster prevention. The then Secretary General of the UN General Assembly, Javier Pérez de Cuéllar of Peru, in collaboration with the National Academy of Sciences of the United States, convened an official meeting of internationally renowned scientists, engineers and representatives of UN agencies in March 1988. As a result of this meeting, the Secretary General decided to create a panel of 25 international experts under the leadership of Frank Press, an influential U.S. seismologist, former director of the National Academy of Sciences and science advisor to President Carter (for the importance of the period in framing the global environmental issue, see the Introduction to this book). This committee met four times between July 1988 and April 1989 and drafted the Tokyo Declaration that launched the first international initiative on the subject, the International Decade for Natural Disaster Reduction (IDNDR). The Tokyo Declaration emphasizes two aspects: the power of science and technology in reducing the damage caused by "natural" disasters and the importance of educating people. Faith in scientific progress, thus, influenced the birth of the first international initiative on disaster reduction, as evidenced by the letter from Frank Press to Javier Pérez de Cuéllar when the expert report was delivered in 1989:

> Fatalism is no longer an acceptable attitude; it is time to use all the resources of scientific and technical progress to reduce the human tragedies and economic losses resulting from natural disasters.

Three imperatives were put forward to justify the need for this International Decade: material motivations linked to the fact that disasters are both causes and products of underdevelopment, the strong emotional and moral implications of the decade and the challenge of reducing human suffering, and finally the intellectual challenge that this project represents for the researchers concerned. A resolution proclaiming the opening of the Decade was proposed and brought before the United Nations General Assembly in December 1989.

The struggle between the two scientific framings of disasters – the hazard-centered framing and the vulnerability-based framing – continued throughout the Decade. Unsurprisingly, the early years were strongly marked by an emphasis on physical vulnerability reduction measures: building reinforcement, knowledge of materials resistance and participation in urban conferences played a central role in the activities of the early 1990s. Research that improves hazard prediction, especially earthquakes and volcanic eruptions, is also highly encouraged, allowing scientists involved in the Decade to fund their research and increase their prestige (Hannigan 2012; Cabane and Revet 2015). When the first international conference on natural disaster reduction was held in Yokohama, Japan, in May 1994, it was time to take stock of the Decade's activities. This was the opportunity for some social scientists to launch the first salvo of criticism. Ian Davis and Mary Myers, actively involved in research on vulnerability factors, published their "observations" of the conference in the journal *Disasters* (Davis and Myers 1994). They point out that many participants at the Yokohama conference expressed disappointment. They regretted the weight given to the sole scientific and technological framing of the conference, despite the notable presence of different voices, notably during a session organized by the Red Cross and UNICEF on "vulnerable communities", which highlighted traditional warning systems or local and family strategies put in place by inhabitants to cope with risks.

Nevertheless, the Decade will give rise to a certain number of international or regional initiatives that will encourage meetings, circulations and gatherings of various researchers and actors. For some networks, the Decade is a godsend that will allow them to make their work and their analyses of risks and disasters known. Actors are emerging in certain regions of the South, in Latin America, South Africa and Asia, who have a different vision of disasters than the one that led to the Decade. These figures, often social science researchers specializing in developing countries, heavily involved in applied programs and whose networks extend into the world of NGOs and development as well as public policy, are gradually succeeding in imposing a new framing in international arenas than that of the Earth sciences (Cabane and Revet 2015). The framing using vulnerability, carried by the social sciences, leads to the notion of "disaster risk reduction" (DRR), which is broader than prevention alone. NGOs, which had relatively little presence in the international arena at the beginning of the Decade, also took on an increasingly important role in the 1990s, in parallel with their growing role in the development world.

5.7. Friction between climate and disaster risk reduction framing

In international arenas, since the 2010s, we have seen increasing encouragement to integrate the theme of "climate change" and disaster risk reduction. This can be done through incentives to set up programs in this direction, as the World Bank is doing with the Global Facility for Disaster Reduction and Recovery (GFDRR) by encouraging climate change adaptation projects to act on disaster risk reduction, or through a multitude of initiatives (Hannigan 2012).

As John Hannigan (2012) shows, the link between disaster risk reduction and adaptation to climate change requires first of all the development of a discourse. This is based on figures, and in particular on figures produced by international mechanisms such as the database of the Centre for Research on the Epidemiology of Disasters at the University of Leuven, EM-Dat, one of the mechanisms that helped produce the international world of "natural" disasters in the 1990s (Revet 2018). These figures make it possible to establish that disasters related to climatic events are those that affect the greatest number of people, on a global scale, and those that cause the most damage, even they are not the most deadly, as earthquakes remain the hazards that produce the most deadly disasters.

In practice, the link between climate change and disaster risk reduction is made primarily through the notions of adaptation and resilience and not through the issue of reducing greenhouse gas emissions. In the context of adaptation and resilience, it is a question of taking note of global warming and the risks it entails, and of acting to reduce the harmful consequences, of putting in place, often on a local or even individual scale, measures that enable coping and recovering. One might think that in this perspective, the most reflexive paradigm would be the one that emphasizes the reduction of structural factors of vulnerability. However, as a closer analysis of projects implemented in the world of natural disasters that are based on the notion of resilience shows (Revet 2018), the tendency is rather toward the implementation of small-scale projects, which aim not to reduce structural vulnerability to destructive events – which the field of emissions reduction could do in particular – but to confront them and recover from them as best as possible. While a reading in terms of global change would *a priori* impose global and structural policies aiming at reducing the causes of these changes and their harmful effects, the projects implemented are most of the time on the scale of a district, a village or a community and are rarely structural, calling for adaptation to these new risk situations.

Moreover, in the international disaster world, some regret that the "climate turn" – which has consisted in a focus of resources and projects around this theme, including in the world of "natural" disasters historically dominated by the geomorphological issue, disasters related to earth movements – has resulted in a refocusing of attention on a single cause, instead of the multicausal analysis that prevails with the attention paid to vulnerability factors (for illustrations and a discussion of this tension, see Part 1). As one such researcher argued back in 2009:

> One big thing is whether this climate change emphasis will push us back to a mono hazard approach instead of multi hazard approach, attentive to the diversity of multi social conditions and factors. (Interview, Geneva, May 2009)

These critical researchers also fear that this focus on "extreme events", which climate change is making more frequent, will result in even greater attention being paid to major and exceptional events, and consequently, will lead to anticipatory measures:

> So you are back to response and early warning systems, early warning, early warning which doesn't... It resolves the singular problem of saving lives and it doesn't resolve... (Interview, Geneva, May 2009)

This focus on major events and on warning measures is to the detriment of smaller, less visible and more routine disasters, such as fires in informal settlements or floods and landslides, which change the conditions of vulnerability of societies and social groups and require risk reduction actions rather than just emergency management and anticipation.

While some have been advocating for years for the "integration" or "inclusion" of climate change adaptation in disaster risk reduction projects (Kelman *et al.* 2017), it is by claiming an approach based on vulnerability factors, and the primacy of research on risk reduction over that on climate change. Moreover, this integration of the two themes, while it may be scientifically relevant, is far from being facilitated by an institutional organization of two very distinct worlds, that of risk reduction and that of climate change adaptation, with differing histories and relationships to expertise (Buffet and Revet 2017).

5.8. Conclusion

Whether "natural" risks and disasters are considered as the manifestation of exceptional natural phenomena or, as is increasingly the case today, as the consequence of anthropic actions that have increased the conditions of vulnerability of the societies exposed to them, naturalist thinking, which assumes a separation between nature and society, remains the framework of thought most often used by professionals in the international world of "natural" disasters. Neither the shift toward vulnerability in the 1970s, which "denaturalized disasters", nor the climate change of the early 2000s, which highlighted a new kind of risk, more global in nature and with clashing time scales, have fundamentally disrupted this naturalist vision of disasters. Very few discourses or practices in this professional world concretely consider the interdependence of systems, the interrelations between humans and non-humans, and reciprocities other than from the perspective of "ecosystem services" or the benefits rendered by nature to humans.

It is necessary to move away from this professional world in order to find, among those locally engaged in a search for meaning in the disasters they experience, an ability to deal with different interpretations of disasters, whether naturalistic, animistic, religious or expert (Revet 2010), and in so doing to combine measures of all kinds, which do not rely solely on the separation between humans and their environment, but engage many other types of relationships.

However, we can measure the limits of this naturalist thinking to take the measure, understand and imagine the answers to the catastrophes that are shaking the planet and its inhabitants. Let us take as an example the catastrophe caused by the earthquake and tsunami in Tōhoku, Japan, in 2011, which led to the Fukushima nuclear disaster, for which it is very difficult to disentangle the natural from the technological issues or to separate the human consequences from those on the lives of wild boars and mushrooms and their relationships (Houdart and Pavy 2019), or the one caused by the global spread of the COVID-19 pandemic in 2020 that questions things as diverse as how we treat the animals we farm for food, the financing of our health systems, or the impact of deforestation on wildlife habitats; it becomes clear that naturalistic thinking is no longer of much use to us. We come up against its limits in silos, in sectors, in broad categories.

"Is the COVID-19 crisis a 'natural' disaster?" is probably one of the questions I was asked most often while confined in the early months of 2020, to see if I was likely, from my particular expertise, to say anything about it. It is impossible to give a reassuring yes/no answer to this question. The multiple interactions between a virus, animals, airplanes, humans, respirators and masks clearly prevent us from putting this disaster, like all others, in a precise and reassuring box.

The institutions of the international "natural" disaster world all say something about COVID-19[10]. They also had a lot to say about the Tōhoku earthquake and tsunami in 2011, but found themselves rather silent on the Fukushima nuclear disaster, demonstrating a way of thinking that attempted to separate the natural world from the earthquake from the nuclear disaster of Fukushima. What these institutions say and do only barely manages to break out of this way of thinking, which clashes with the circulations, connections and barrier crossings that challenge us, and that these "hyperobjects" (Morton 2013) bring to light.

In this context, does the notion of the Anthropocene allow us to take another look at what is happening to us? A look more inclined to go beyond the separate vision of naturalism? By presenting an "Earth system" in which humans are both a major transforming force from now on, but also one element among others, the notion has the virtue of attempting to go beyond the purely naturalistic vision of the relations between humans and non-humans and to show us "an intricate network in which 'social' and 'natural' arrangements are reproduced mutually by a thousand channels" (Fressoz and Bonneuil 2013). Its limitations undoubtedly lie in its difficulty in capturing humanity – the anthropos – in its diversity, which authors have invited us to think about by contrasting it with several alternatives: Capitalocene, Plantationocene (Haraway 2014; Haraway *et al.* 2016), or even Negrocene (Ferdinand 2019), with the idea of reintroducing the diversity of situations, inequalities and power relations that have historically traversed the

10 See, for example, the large number of articles on COVID-19 at: www.preventionweb.net/english/.

human actions at the origin of this change in the geological era. Whatever the term used, the potential lies in the designation of interrelationships, interdependencies, networks and in the attempt to go beyond the strict naturalistic separation. It would then be one more step on the path of reflexivity, to understand, in their inequalities, the entanglements produced between locally and historically situated humans and the Earth system. If the international world of disasters is looking for a way out to think about the complexity of relationships, perhaps it will have to include in its next reports a reflection on what disasters mean during the Anthropocene moment?

5.9. References

Barth, K.-H. (2003). The politics of seismology: Nuclear testing, arms control, and the transformation of a discipline. *Social Studies of Science*, 33(5), 743–781.

Blaikie, P., Cannon, T., Davis, I., Wisner, B. (1994). *At Risk: Natural Hazards, People's Vulnerability and Disasters*. Routledge, London.

Buffet, C. and Revet, S. (2017). UN institutions doing disaster risk reduction and climate change adaptation: UNISDR, UNFCCC, and IPCC. In *The Routledge Handbook of Disaster Risk Reduction Including Climate Change Adaptation*, Kelman, I., Mercer, J., Gaillard, J.-C. (eds). Routledge, New York.

Burton, I., Kates, R.W., White, G.F. (1968). The human ecology of extreme geophysical events. *FMHI Publications*, Paper 78 [Online]. Available at: http://scholarcommons.usf.edu/fmhi_pub/78.

Burton, I., Kates, R.W., White, G.F. (1978). *The Environment as Hazard*. Oxford University Press, New York.

Cabane, L. and Revet, S. (2015). La cause des catastrophes. Concurrences scientifiques et actions politiques dans un monde transnational. *Politix*, 111(28), 47–67.

Carr, L.J. (1932). Disaster and the sequence-pattern concept of social change. *The American Journal of Sociology*, 38(2), 207–218.

Davis, I. and Myers, M. (1994). Observations on the Yokohama World Conference on natural disaster reduction, 23–27 May 1994. *Disasters*, 18(4), 368–372.

Descola, P. (2005). *Par-delà nature et culture*. Gallimard, Paris.

Descola, P. (ed.) (2018). *Les Natures en question. Colloque de rentrée du Collège de France 2017*. Odile Jacob, Paris.

Doel, R.E. (2003). Constituting the postwar earth sciences: The military's influence on the environmental sciences in the USA after 1945. *Social Studies of Science*, 33(5), 635–666.

Edwards, P.N. (2010). *A Vast Machine. Computer Models, Climate Data, and the Politics of Global Warming*. MIT Press, Cambridge.

Felli, R. (2016). *La grande adaptation. Climat, capitalisme et catastrophe*. Le Seuil, Paris.

Ferdinand, M. (2019). *Une écologie décoloniale – Penser l'écologie depuis le monde caribéen*. Le Seuil, Paris.

Fressoz, J.-B. and Bonneuil, C. (2013). *L'Événement Anthropocène. La Terre, l'histoire et nous*. Le Seuil, Paris.

Fritz, C.E. and Marks, E.S. (1954). The NORC studies of human behavior in disaster. *Journal of Social Issues*, 10(3), 26–41.

Hamblin, J.D. (2013). *Arming Mother Nature. The Birth of Catastrophic Environmentalism*. Oxford University Press, New York.

Hannigan, J. (2012). *Disasters without Borders: The International Politics of Natural Disasters*. Polity, Cambridge.

Haraway, D. (2014). Anthropocene, Capitalocene, Chthulucene: Staying with the Trouble. *Arts of Living on a Damaged Planet*, University of California, Santa Cruz. Available at: http://opentranscripts.org/sources/anthropocene-arts-of-living-on-a-damaged-planet/ [Accessed 31 August 2021].

Haraway, D., Ishikawa, N., Gilbert, S.F., Olwig, F., Tsing, A.L., Bubandt, N. (2016). Anthropologists are talking – About the Anthropocene. *Ethnos, Journal of Anthropology*, 81(3), 535–564.

Hewitt, K. (ed.) (1983). *Interpretation of Calamities*. Allen and Unwin Inc., Boston.

Houdart, S. and Pavy, M. (2019). On sort donc les tripes petit à petit. Consistance d'un territoire contaminé [récit]. *Terrain. Anthropologie & sciences humaines*, 71, 116–139.

Kelman, I., Mercer, J., Gaillard, J.-C. (2017). *The Routledge Handbook of Disaster Risk Reduction Including Climate Change*. Routledge, New York.

Langumier, J. (2008). *Survivre à l'inondation : pour une ethnologie de la catastrophe*. ENS Éditions, Lyon.

Manzo, K. (2010). Imaging vulnerability: The iconography of climate change. *Area*, 42(1), 96–107.

Morton, T. (2013). *Hyperobjects: Philosophy and Ecology after the End of the World*. University of Minnesota Press, Minneapolis.

Prince, S.H. (1920). Catastrophe and social change: Based upon a sociological study of the Halifax disaster. *Studies in History, Economics and Public Law*, 94, 1–152.

Quarantelli, E.L. (1954). The nature and conditions of panic. *The American Journal of Sociology*, 60(3), 267–275.

Schemper, L. (2016). La prévention des catastrophes naturelles et les organisations internationales du temps de la SDN au lendemain de la guerre froide. Quelle place pour l'environnement ? *Études internationales*, 47(1), 29–55.

Sorokin, P.A. (1942). *Man and Society in Calamity: The Effects of War, Revolution, Famine, Pestilence upon Human Mind, Behavior, Social Organization and Cultural Life*. E.P. Dutton, New York.

Revet, S. (2010). Le sens du désastre. *Terrain*, 54, 42–55.

Revet, S. (2018). *Les coulisses du monde des catastrophes "naturelles"*. Éditions de la FMSH, Paris.

Revet, S. (2020). *Disasterland: An Ethnography of the International Disaster Community*. Palgrave Macmillan, Basingstoke.

White, G.F. (1945). Human adjustment to floods: A geographical approach to the flood problem in the United States. Research document no. 29, University of Chicago, Chicago.

Wisner, B., O'Keefe, P., Westgate, K. (1977). Global systems and local disasters: The untapped power of peoples' science. *Disasters*, 1(1), 47–57.

6

The Difficult Birth of the Risk Society and the Relegation of Social Sciences

Pascale METZGER[1] and Julien REBOTIER[2]

[1] PRODIG, IRD, Paris, France

[2] TREE, CNRS – E2S UPPA, Bayonne, France

6.1. Introduction

The risk society[1] refers to a reading of the evolution of the social world through the prism of risk. Despite the question of risk long being used to interpret the social world and modes of governance (Ewald and Kessler 2000), a new grand narrative is unfolding, that of the advent of a reflexive modernity that takes seriously the damaging effects of science and technology, and that leads to new forms of politics. However, there is a certain inability of the social sciences to formulate the challenges posed by risks and the environment in political terms. The framing of the notion of the Anthropocene contributes to reinforcing a naturalist and technical vision that removes or marginalizes social and political approaches (see Chapters 5 and 8). This is the paradox of the risk society, which identifies the mutations of modernity and its implications through risks, without seriously advancing recognition of the social and political dimensions of risk issues, the latter remaining fundamentally reduced to the physical.

---

1 The use of risk society without italics refers to the historical moment of contemporary societies grappling with risk. *The Risk Society*, in italics, refers to Beck's essay (1992).

Risks and the Anthropocene,
coordinated by Julien REBOTIER. © ISTE Ltd 2021.

Starting from a reflection on the risk society, this chapter aims to highlight the objective marginalization of the social sciences and their difficulties in thinking about risk in an autonomous way. The proliferation of scientific articles and books on risk echoes the proliferation of risks that seem to contaminate the social world (Bourg *et al.* 2013). Inserted in a nebulous array of contradictory positions, the term risk appears simultaneously as the object of three distinct worlds: social, material and scientific (Metzger 2017). A critical but not very audible discussion of the Anthropocene (see the Introduction to this book) fades behind a naturalist framing that places material objects and physical phenomena at the heart of risk and environmental issues. The social sciences consider physical and material things from the point of view of social, political and technical constructions. They use concepts that are vague and unstable and are always being rediscussed (Becerra 2012), which systematically contributes to distancing their problems to the detriment of the more legitimate hard science approaches. This is a first epistemological obstacle, which has to do with the objects, issues and methods of different disciplines. In addition, the system of constraints and the relationships of domination in force within the scientific field pose the problem of the autonomy of the social sciences. The production of social science research is increasingly constrained by the institutional order that directs and finances research, in addition to requiring multi- or interdisciplinary research. The notion of the Anthropocene is rooted in the hard sciences and tends to reinforce this trend. In this framework, a significant production of social sciences implies a form of rebellion against social demand and interdisciplinarity dominated by hard sciences in order to resist relegation.

In order to understand the difficult position of social sciences around risks and the environment during the Anthropocene moment, we will first unfold the ambiguities of the *Risk Society* (Beck 1992), seen as risk-taking of a world engaged on the path of reflexive modernity. The following section will explain the contributions of the social sciences that both confirm and criticize the book's theses. The last section will focus on the social conditions of research production and the institutional difficulties that social sciences encounter in thinking about risk in an autonomous way and in making their voice count.

6.2. *The Risk Society*, an ambiguous grand narrative

Beck's work (1986, translated into English in 1992) is an incontrovertible reference in research on risk, but also on modernity, globalization and cosmopolitanism. However, without a definition of risk, the subject is immediately placed at a distance from the academic approach (Chauvel and Ramaux 2002). As an attempt at prospective social theory rather than scientific demonstration, its methodological ambiguity (Pierret 2008) makes its status as a scientific reference all the more paradoxical. However, the text echoes the environmental concerns that are disturbing/disrupting the scientific sphere, the social world and the global public space, and it opens up many avenues for reflection.

6.2.1. *A prospective essay on social theory*

The author starts from the decline of modern industrial society marked by the proliferation of risks, the end of the great structuring narratives and the ideology of progress, and the recomposition of the great social institutions (family, State, work). Through an analysis of the evolution of these social structures, of individualization and of new forms of mobilization, Beck puts us on the "path to another modernity": the emerging risk society.

6.2.1.1. *Proliferation and reconfiguration of risks*

Beck's starting point is the observation that modern industrial societies are confronted with ever greater risks that are less and less controllable, some of which are irreversible and often invisible. The argument is illustrated by the global and environmental threats, technological dangers, nuclear disasters and public health controversies that marked the years 1970–1980 (Bhopal, Seveso, GMOs, contaminated blood, oil spills, acid rain, etc.). The proliferation of risks resulting from the industrial modernity of the 19th century makes it impossible to attribute dangers to external causes. The risks are no longer in nature but in development. The production of wealth is systematically correlated with the social production of risks.

Risks are globalized: "abolish borders" and "the nation-state is being surpassed". They are often invisible, in the sense that their visibility is mediated by science and technology, via an arsenal of measures and expertise, which allow them to "reach scientific existence" (Beck 1992). These risks are not only introduced into the heart of the social and political world, but also into people's daily lives and intimacy, by becoming "the blind passengers of normal consumption" (Beck 1992).

According to Beck, it is impossible to fight against these risks, which generate fears and anxieties throughout society. On the one hand, prevention policies rely on contested experts and scientists, whose knowledge is mainly disciplinary, so necessarily limited and uncertain. On the other hand, it is no longer possible to protect oneself against globalized risks that have become uncontrollable within national political frameworks. Finally, attempts to control risk in turn generate new risks.

6.2.1.2. *Science and technology in question*

Analysis of the role of science and knowledge is considered a major contribution of the *Risk Society* (Bonneuil 2002). Beck emphasizes a rupture in modernity. While it was carried by the central values of progress and rationality embodied by science and technology, it appears that modernity does not only produce wealth but also increasingly uncontrollable risks. Scientific and technological advances produce all sorts of threats and dangers, direct and induced, visible and invisible, whose scope today directly echoes the Anthropocene. For all that, the promises of development (an end to poverty, inequality, famine, etc.) have not been kept (Dubar 2002).

The destructive consequences of modernization, especially in the health field, become public affairs. Controversy takes hold of the whole of society in order to challenge the legitimacy and authority of knowledge and democratic institutions of regulation (Bonneuil 2002). Under these conditions, society "can no longer modernize by relying on rationality and progress, which produce risks; it must no longer delegate implementation of this modernization to researchers, industrialists and financiers" (Dubar 2002).

Today, the status of science is ambiguous in several ways. Although the practice of science plays a crucial role in the production of risks, it is also expected to invent solutions, in the paradoxical context of a questioning of the "modern religion of progress".

6.2.1.3. *Disqualification and rearrangement of the politics*

In the face of the powerful alliance between science, technology and capitalism, Beck observes a progressive diminution of the role of politics, or even its paralysis. Political choices and future options are less the domain of parliament or political parties than of research laboratories and boards of directors (Beck 1992). Risks are "produced by industry, externalized economically, individualized legally, legitimized scientifically and minimized politically" (Vandenberghe 2001). Science and technology play a decisive role in the retreat of politics, because under the guise of neutrality and objectivity, technocracy and rationality carry an invisible political project. Political institutions only administer an uncontrolled evolution of society, while decisions are taken elsewhere (Murard 2002), according to the higher principles of rationality, efficiency and profit.

Faced with the power of global, omnipresent, eminently destructive risks, the product of complex causal chains (Pierret 2008), it is the welfare state that is called into question. The probabilistic calculation and the insurance management of the risks on which it is based (Ewald and Kessler 2000; Castel 2009) are also being undermined, further undermining the credibility of politics by diluting responsibilities. At the same time, new products are appearing on the market via insurance solutions adapted to new situations of growing uncertainty. A new configuration of politics is gradually emerging. The "reflexive society" is the one that will be able to put an end to "the monopoly of knowledge by scientists and experts" and to "the opacity of decision-making processes", as mentioned in Beck's *Risk Society*, which brings science into politics.

6.2.1.4. *Restructuring of the social*

But Beck's project intervenes in a social world transformed by a growing individualization (Murard 2002), which increases uncertainties and risks by destructuring institutions and social forms resulting from industrial modernity. This is the case for the family, gender relations and the labor market, while new social links and roles appear (Dubar 2002).

In parallel with the depoliticization of politics concealed in and by technical rationality, new forms of mobilization are appearing around the awareness of risks. New collectives (environmental movements, citizen participation, civil society associations) are being formed outside the formal game of politics and no longer owe anything to the categories of industrial society (Pierret 2008). Risks give rise to "objective communities of threat" that cut across social categories and eliminate the evidence of socially differentiated exposure and vulnerability.

Between the globalization of risks, the disqualification of social classes and national identities, the end of the role of the State and of traditional social structures, politics captured by the private sector and science, it is the panorama of a form of social cataclysm that is in question.

6.2.2. *The ambiguities of the* Risk Society

Beck's political prophecy and the analysis on which it is based are far from being unanimously accepted (Vandenberghe 2001; Chauvel and Ramaux 2002; Murard 2002). Despite the major contributions to the understanding of the contemporary world, the message conveyed by the *Risk Society* appears paradoxical and liberal.

6.2.2.1. *An eminently paradoxical statement*

Beck points to a form of depoliticization of risk, evident in the critique of expertise (Boudia and Demortain 2014; Henry *et al.* 2015; Reghezza 2015), which needs to be repoliticized through the advent of the risk society. Yet to respond to risk, there is a demand for more science that produces more risk, intensifying the call for more science. The criticism of science and progress carried by reflexivity becomes the support of a form of "expansion of science" (Bonneuil 2002). If science produces risks by revealing them, it also contributes through "risk corrections" to the appearance of new risks, in other fields, other places or other times (Metzger and D'Ercole 2011). From this point of view, more science leads to more risks, which lead to more science and not more politics, even though science is a form of exercising politics.

The other great paradox is the critique of the logic of capitalism carried by the *Risk Society*, at the same time as the affirmation of its inevitable expansion. Since risk and security constitute an infinite market horizon (Bonneuil 2002), capitalism is wrapped in the clothes of sustainable development and risk management. Based on this observation, Beck postulates a general and irreversible movement of technical and economic globalization, that of "a rallying to the inevitability of capitalism" (Vibert 2012). Class structures then fade away in a society swept along by this inevitable movement, where the engine of history becomes fear. Conflicts over the sharing of wealth give way to conflicts over the management and distribution of risks (Dubar 2002). The challenge is no longer to produce security, but to adjust the state of insecurity, to live – unequally – with risk. This form of abdication in the face of the inescapable is reminiscent of the criticisms leveled at the notion of resilience (Rufat 2011; Quenault 2013; Reghezza 2015) and at a liberal political agenda devoid of social discernment. However, if the search for profit and the most advantageous combination of capital and labor remains the driving force of a globalized competition that subjugates the state (Beck 1992), the logics of capitalism are attacking historically acquired social rights (Castel 2009).

6.2.2.2. *A fundamentally liberal approach*

By advocating for the emergence of individuals and by underlining the incapacity of States, the *Risk Society* is unquestionably liberal. In this classless and universal society, the emergence of a reflexive subject disqualifies collective thoughts of the social. More than the end of capitalism, the *Risk Society* announces the "peremptory liquidation of the proletariat as a political subject" (Murard 2002), a discourse typical of liberal sociology on the end of social classes (shared enrichment, accelerated mobility, dissolution of cultural referents, reduction of social and political consciousness; Chauvel and Ramaux 2002). Beck identifies a capitalism that retains its structures and its problems of social inequality at the same time as the emergence of individuals liberated from social structures and facing a common future (Vandenberghe 2001). Without social class, without democratic citizenship, without regulation by the state which no longer has control, individualization becomes the only horizon (Chauvel and Ramaux 2002). In this framework, politics is embodied by new forms of collective action that no longer owe anything to inherited social institutions. According to Beck, the risk society leads to the emergence of a reflexive and autonomous subject. The individual becomes a form of paragon of liberal ideas, replacing the thoughts of the social which are based on collectives, institutions and socially constructed regulations.

Beck's essay takes note of the fact that states ratify market mechanisms as the dominant form of social regulation. The powerlessness of states to deal with environmental issues is verified by the failures of global negotiations, the power of the economic actors of globalization, the continuing race for progress and innovation and the permanence of risk-producing activities. Within nations, the welfare state is retreating, which is unable to resist globalization. Reforms are multiplying, all tending to reduce coverage of social risks (health, retirement, unemployment) and to reduce the scope of public services (Ewald and Kessler 2000; Castel 2009). The de-socialization of the state stems from its submission to the capitalist economy and the arguments of scientific rationality that legitimize public action. Technocratic government endangers the social state as well as democracy (Vandenberghe 2001) and is the basis for a rhetoric that postulates the absence of alternatives in the context of neoliberal globalization: *there is no alternative*.

6.2.2.3. *Forward thinking to be updated*

Nearly 40 years after its publication, at a time of climate change, environmental concerns and the success of collapsology, we can ask ourselves what is happening to the risk society, understood as the advent of a reflexive society that takes seriously the risks it produces, and that renews the democracy that has been damaged by technocracy.

First, we acknowledge the fact that risk is a relevant point of entry for understanding the social world. In the Anthropocene moment, the generalized agenda of risk issues is characteristic of contemporary societies. However, we observe the intensification of a management of the world that is both liberal and technocratic. The multiplication of social movements, simultaneously local and global, does not translate into a democratic deepening of risk management, which remains formulated as a primarily technical and scientific issue (see Chapter 5). The opening up to other types of knowledge is rhetorical, since the latter do not question the authority of scientific discourse, but rather strengthen its legitimacy and relay its dissemination in the social world.

On the other hand, the *Risk Society* refers rather to an egalitarian society, involving "communities of objective threats". This point of view is reminiscent of a naive idea of the Anthropocene according to which "we" (humans) are "all in the same boat". However, if certain threats are indeed likely to affect "everyone", the forms of exposure, the means of protecting oneself from them, of coping with them or of living with them are far from being the same for everyone. Risks retain their discriminating character. In addition to the ultimate value of freedom and the pressing demands for security, the cardinal notion of justice recognizes the irreducible differentiations of the world (Smith 2010). The Anthropocene moment is not immune to these tensions, in complete contradiction to Beck's egalitarian vision.

6.3. Contrasting contributions of the social sciences

The science and technology that produce risk and modernity, to which the *Risk Society* refers, seem to refer only to the field of hard sciences. One can then wonder about the contribution of the social sciences to the problematic of risks as well as to the knowledge of the social world. Analyses of the social sciences confirm the thesis of the *Risk Society* at the same time as they complexify it, contest it, and even contradict it.

6.3.1. *The evolution of the relationship to risk*

6.3.1.1. *From the illusion of control to the recognition of uncertainty*

In France, the problem of risk entered the social sciences in the 1980s through disasters and the notion of "major technological risk" formalized in *Civilisation du risque* (Lagadec 1981). This work is based on analysis of catastrophic events of the 1960s and 1970s from the perspective of the vulnerability of technical systems. We find the arguments in the *Risk Society*: the novelty of dangers caused by technological and organizational choices, uncontrollable impacts in space and time, the inability of security measures to control the danger, economic, political and societal consequences, the questioning of democracy, calling for an awakening of collective responsibility and the opening of technological decision-making to citizens.

A second important work is *La société vulnérable* (Theys and Fabiani 1987), which deals jointly with risks and the environment from the point of view of research and management. It points out that the vulnerability of society reflects urbanization, the concentration of activities and the development of increasingly complex and dangerous technologies, but also the improved performance of technical systems and growing security.

Twenty-five years after *La société vulnérable*, *Du risque à la menace* (Bourg *et al.* 2013) challenges the idea of a reflexive society, concerned with limiting the secondary and dangerous effects of techno-scientific developments: insofar as "risk and its methods of control have become producers of threat", research and action accompany dangers rather than limit them. The book questions what approaches in terms of risk – understood as a danger that can be taken in hand, calculated and controlled – make visible or hide. Henceforth, more than risk, "threat" encompasses a whole range of phenomena, from exceptional "natural" events to economic or political threats, which are likely to follow one another in an uncontrollable way, leading to the idea of "major risk" (Lagadec 1981). The threat – seen as a set of imbricated dangers impossible to control – no longer enters into the "putting at risk" (Ewald 1986), which inscribes the dangers in management procedures. It escapes the forms of control and reduction of uncertainty specific to risk.

This shift in social science approaches from risk to threat reflects a rise in uncertainty and refers to what may characterize the Anthropocene (Reghezza 2015). It is within this framework that we are witnessing the substitution of vulnerability by resilience.

6.3.1.2. *From vulnerability to resilience*

Vulnerability is a key notion in the social sciences. Initially, it referred to a certain fragility and susceptibility to damage, resulting from exposure to hazards such as social, economic, political, technical conditions (Quenault *et al.* 2011; Becerra 2012). Subsequently, the notion of vulnerability has been extended to the capacities of social groups in the face of risks. Vulnerability is no longer the consequence of risk on a passive social world but the ability of the social world to influence the risk. Vulnerability is concretely embodied in the differentiated consequences of the impact of a hazard (external to the social world) on individuals, social groups, societies or territories (Blaikie *et al.* 1994).

However, vulnerability is almost always posited in the face of a hazard – seen as a phenomenon outside the social world. This idea of the exteriority of the threat is in contradiction with the *Risk Society*. A rare definition that does not refer to the hazard is the following: "Social vulnerability derives from the activities and circumstances of everyday life or its transformation" (Hewitt 1997; cited by Quenault *et al.* 2011). This formulation makes it possible to think of vulnerability as being anchored in the normal functioning of the social world, free of the hazard and the disaster. An unprecedented break occurs in the vision of vulnerability but also of risk itself.

Since the 2000s, the term "resilience" has spread in the social sciences and public policy, as an echo of Beck's liberal and individualist *Risk Society*. Resilience expresses the ability to absorb shocks and return to normal functioning after any kind of disruption (Quenault 2013). In the field of risk, resilience has come to compete with vulnerability as a universal solution. Despite the many criticisms it has received (Rufat 2011; Bourcart 2015), resilience is gaining ground by evoking the possibility of an epistemological break with technicist visions of risk and the environment (Djament-Tran and Reghezza-Zitt 2012).

The substitution of resilience, which echoes the liberal vision of the world, for vulnerability, which is the bearer of a critique of development, is a subjugation of ways of thinking in the social sciences. Resilience participates in a form of acceptance of risks and crises for which the role of political actors becomes secondary (Petit *et al.* 2014; Quenault 2015). The responsibility for dealing with crises is redistributed among individuals whose behaviors, scrutinized, serve as techniques for governing (Borraz 2013; Felli 2014). Under the guise of an agreed critique of technocratic and top-down approaches, of the promotion of the autonomy of communities and individuals, and of their own capacities to "cope" with risks and crises, to "educate themselves" and "adapt" to "new threats" (Revet 2011), resilience clears States and public authorities of their modern role in the organization of the social world (see Chapter 4).

The widespread dissemination of resilience as a universal solution acknowledges the failure of critics of the notion, who were unable to create a way of thinking about risk and its management that did not ignore the political project. Political science analyzes risk as a "technique of government [...] that contributes to reconfiguring the social and the political" (Borraz 2013). Risk is thus anchored in the field of the social and the political, the result of a work of construction. In this, social science research documents processes of risk construction and social organization in which science and technology count, but among many other elements. With what results?

6.3.2. *Unquestionable but little heard contributions to risk management*

6.3.2.1. *Three meanings of social construction*

If the idea of the "social construction" of risk federates social science approaches to risk, it is appropriate to identify at least three conceptions. The first refers to the "concrete" production of risks by society and (bad) development (Theys and Fabiani 1987; Beck 1992). A radically opposed conception considers "risk-taking" or the "risk factory" (Ewald 1986; Gilbert 2003), referring to all the social processes, investments and actor alignments that lead to the formulation of a problem in terms of a calculable risk that can be managed within the framework of public policies (Gilbert and Henry 2009). This conception refers to something equally real, which is played out within the social world, about things that engage material danger.

Quite different is the statement that risk "does not exist" (Kermisch 2012). Behind this abrupt formulation, we find, on the one hand, the idea that risk is always potential, probable and calculated, and as such it does not exist. When the risk "comes true", when it becomes "real", it is called "accident", "crisis" or "catastrophe". On the other hand, there are approaches based on representations or perceptions that question the objective reality of risk by mobilizing sociological, psychological and anthropological approaches (Peretti-Watel 2000). Risk is seen as a belief, a perception, a representation in the line of culturalist and behaviorist analyses. If it is subjective, its reality and its objective condition are evacuated. The hard sciences are particularly receptive to analyses of risk perception, which reinforce their position: since social perceptions have no material and objective reality, they do not exist. The only legitimate reality of risk is based on the physical approach to damaging phenomena, against which any measure is taken.

6.3.2.2. *The status of the event: an epistemological challenge*

Most works on risk take the disaster directly or indirectly as a starting point or as a perspective, or both. As an event, the catastrophe, past or probable, is at the heart of risk. It can be situated in the order of the extraordinary, which delimits a before and an after: "The risk and its realization thus refer once again to an exceptional phenomenon" (Bourg *et al.* 2013). Now, "for the social sciences, the event poses a problem" (Bensa and Fassin 2002) in the sense that their aim is to identify forms of "laws" and "regularities" of the social that the event seems to suspend (Morin 1972). Hence, the epistemological necessity for the social sciences to reposition the event in the regularities of the social. This is done, on the one hand, through analyses that identify regularities in the unfolding of crises (Lagadec 1981; Godard *et al.* 2002), and, on the other hand, through efforts to reinscribe risk and crises in the normality of political and social processes (Ewald 1986; Hewitt 1997; Gilbert and Lascoumes 2003; Gilbert 2013).

These approaches posit risk as a form of normality, inscribed in the very modalities of the functioning and development of modern society (Beck 1992; Borraz 2013). In a vision of risk torn between the exceptional and the ordinary, the event is increasingly disqualified. Taking an interest in "what holds the world together" on a daily basis, consistent with a society that constantly produces new risks, opens up research perspectives: thinking about ordinary security far from crisis situations (Gilbert 2016). In this respect, the Anthropocene "event" raises questions if it does not allow us to explain the ordinary mechanics of a social world grappling with new environmental issues.

Whether they are natural, technological or environmental risks, risks are made visible by science, which designates certain objects or phenomena (asbestos, GMOs, acid rain) as dangerous through technical and normative measurement devices. In this case, it is science – here the hard sciences – that designates potential hazards. But for the social sciences, designating risk implies passing through the social world and the controversies held in enlarged arenas (well beyond the perimeter of the hard sciences). It is the risk factory that inscribes the problems in the field of risks (Ewald 1986).

However, the question of risk raises a tension between ignorance and knowledge specific to the scientific field: "Far from reducing ignorance, knowledge increases uncertainty and thus risk" (Luhman quoted by Le Bouter 2014; Henry *et al.* 2015). This position refers to the post-normal science (Ravetz 1992) according to which science does not roll back ignorance, but increases the unknown and the uncertain. The growing tensions between uncertainty and knowledge at the heart of risk issues explain the development of work on expertise and the relationship between science and society (Henry *et al.* 2015). Such a development opens the way to the sociology of science, insofar as the question of risk necessarily implies controversy and skepticism (Gilbert and Henry 2009; Larrère and Larrère 2013). The simple question of whether or not there is danger is the subject of controversy: what is true is no longer decided solely by scientific knowledge, which no longer enjoys the benefit of authoritative argument. We are thus witnessing a form of opening up of the decision to the social world through processes of recognition of lay knowledge, participation or consensus conferences, which constitute new research objects for the social sciences. This is the benefit of an interpretation in terms of "putting at risk" and the "risk factory" (Gilbert 2003), which places resolution of the controversy, always provisional, not in the field of normal science (metrological controversies), but in the contingent field of the social world (axiological controversies).

The social sciences provide information on what constitutes risk, in regards to what key values, and at the end of what process. However, analysis of the conditions of production of knowledge on risks also lifts the veil on the relative positions in which the social sciences are held within the scientific field.

6.4. Social sciences caught in a world of constraints

It is clear that the knowledge provided by the social sciences has a very limited impact on the social world and the formulation of public policies. The social sciences suffer from a lack of legitimacy in risk and environmental issues, especially when compared to the hard sciences. They also suffer from the social conditions of production that increase their position of subordination within the scientific field (Bourdieu 1975).

6.4.1. *The superiority of calculation over discourse*

In the social and scientific world, there is a belief in the superior capacity of the hard sciences to tell the truth, supported by figures, measurements and technical devices. For the "modern technical reason" (Feenberg 2002), calculation and rationality, considered as true, neutral, objective and universal, found a *superior principle of legitimization* (Boltanski and Thévenot 1991; Godard 2004).

Despite developments on the loss of confidence in science and technology, the rhetoric of the figure remains central to demonstration and evaluation (Godard 2004), as shown by the all-out dissemination of the new public management, piloting by indicators or benchmarking, which measure the efficiency of individuals and organizations. The same is true for the evolution of the conditions of production of scientific research (Shanghai ranking, publication scores; Bouleau 2003, p. 88), while "any quantification is a possible extension of the market". In this, science – hard science – and technology hold the instruments of legitimization of advanced capitalist societies.

This is the register of industrial modernity, where calculation is involved in the designation of risk. For experts and public authorities alike, risk can be objectified by a function between the probability of occurrence of a phenomenon and the amount of damage it could cause. This is the classic "hazard × vulnerability" statement that is found both in research and at the heart of prevention policies. However, figures also conceal ignorance and lead to a false reduction of uncertainty. However, the supremacy of quantification and instrumental devices, which are the prerogative of the hard sciences and technology, continues to operate.

In contrast, the knowledge provided by the social sciences is often of the order of interpretative discourse. They situate and explain their object of research in its historical, social, political and territorial context. On the grounds of an exaggerated relativism or a radical constructionism, social sciences are quickly accused of being contingent, subjective, even ideological and dogmatic, especially when they deal with the discovery of relations of domination (see the criticisms made of Bourdieu). Conversely, they are rather well accepted when they produce figures, even though the use of quantification puts objects on an equal footing, disqualifies interpretation and crowds out meaning (Lemoine 2009; Bouleau 2013).

For the last 15 years, calls for projects have systematically insisted on multidisciplinarity and the need to take into account the social dimensions of risk. However, the latter most often consist of a very limited vision of vulnerability (counting elements exposed to a hazard, perception of risks). Such expectations, both on the part of the hard sciences and on the part of managers, pose the risk in terms of awareness and acceptance, and make it possible to work toward the implementation of public policies, far from the clarification of the risk factory.

6.4.2. *Financial conditions and problematic orientations*

In general, research is increasingly dependent on project-based funding in response to institutional demand. In France, the National Research Agency serves the "implementation of project-based research funding" based on a "competitive selection process that respects international standards". Its objective is, of course, to "contribute to the development of science and technology" and to "promote creativity", but also to "target research efforts on economic and societal priorities defined at the highest level of government", "encourage interactions between disciplines" and "intensify public-private links"[2]. The explicit orientation of scientific research is dictated by the objectives and priorities of the State in a competitive and multidisciplinary context (Metzger 2017). In this sense, we are witnessing a loss of autonomy of the scientific field as a whole (Bourdieu 2001), a loss that is even more pronounced for the social sciences because autonomy is also based on a "right of entry". It is a question of the competence incorporated by the discipline for the right to tell the truth. Now "everyone feels entitled to meddle in sociology and to enter the struggle about the legitimate vision of the social world" (Bourdieu 2001). This is not the case when it comes to climate modeling.

Multidisciplinarity does not call into question the existing relationships of domination within the scientific world. On issues of risk and the environment, the hard sciences are considered to be the true holders of knowledge, leaving the social sciences to play the role of *dancer* (Pestre 2004). The hierarchy of disciplines is inscribed in the institutional functioning of research. It is reflected in the asymmetrical involvement of the hard sciences and the social sciences in multidisciplinary projects on risks and the environment, and in the formatting of social demands that impose a naturalist vision of the issues (Metzger 2017).

The institutional demand in terms of research is first and foremost for knowledge of the physical phenomena involved in risks, and thus for knowledge provided (and research driven) by the hard sciences (Metzger 2017). As a result, the social sciences must comply with the dominant conceptual and problematic framework. They are assigned to produce knowledge in "risk areas" or on "vulnerable populations" designated by the hard sciences or useful for the implementation of public policies. They are pushed toward a form of social engineering devoted to the accounting

2 www.agence-nationale-recherche.fr/missions-et-organisation/missions/.

evaluation of vulnerability, resilience, perceptions of risk and acceptability, which is hardly compatible with the analysis of the risk factory.

6.4.3. *Between expectations of operationality and inaudible criticism*

The operationality of research in response to social demand contributes to its legitimacy. However, expecting social sciences to serve a purpose "is always asking them to serve power, whereas [their] scientific function is to understand the social world, starting with power" (Lahire 2004). The approach in social sciences proceeds first of all from a reformulation of the "problems" stated by the social demand (Castel 2009), which does not suit the "applicants" who are exposed to receiving answers to a question they did not ask.

Although the relevance of the discourse on the social dimension of risk is well established, only certain aspects of it are effectively mobilized. Risk managers require an evaluation of the "exposed elements" and the "vulnerability" of the population, which allows for numerical and operational representations in the form of maps and indicators. The accounting evaluation of vulnerability is based on socioeconomic characteristics (on which prevention policies generally do not act) or documents social perceptions and behaviors of the population identified as problematic because of distortions with the "real" risk as described by the hard sciences (i.e., the hazard).

The viewpoint that consists of positing the population as incompetent, irrational and ignorant of risk issues (Joly 2001) excludes taking the population's lay knowledge seriously. At the same time, the diversity of risks that the population must face, but also the necessary arbitration between conflicting interests encapsulated in management policies, is neglected (Dourlens 1991; Metzger and D'Ercole 2011). Focused on what the population says and does "in the face of risk", these social science practices participate in the dissemination of the hard science point of view (Coanus *et al.* 2010; Peretti-Watel 2010), founding "awareness" policies to develop "good practices" and "risk culture" (Revet 2011). Also, social sciences are those who know how to "talk to people". They are called upon to "translate" knowledge of the hazard, to make it more acceptable and to encourage the adoption of good behavior.

But to stick to a critical and radical position based on vulnerability is to ratify the weak operationality of the notion insofar as concrete risk reduction action becomes impossible (Metzger and Robert 2015). Faced with the ever more expected operationality, critical approaches to vulnerability are incapable of proposing transformations of the development modalities that prevention policies could implement. The formal insertion of vulnerability into public policy has neutralized its critical scope by locking it into a quantifiable vision of potential losses, or into a utilitarian input on the perceptions and behaviors of the population.

On the other hand, the interpretation of risk as a new principle of government and structuring of the social world (Borraz 2013; Revet and Langumier 2013) is possible because it is an insight that does not claim direct operational utility. This reading of risk is not intended to diminish risks or to intervene in concrete situations, but to explain them, to decipher their role in the configuration of the social world. In this respect, the social sciences are building an autonomy that is inversely proportional to the echo they have in the public arena, which seems to be confirmed by the Anthropocene moment.

6.4.4. *The imperative framing of the Anthropocene*

For a long time, the social sciences have resisted investing in the environment as an object of research (Henry and Jollivet 1998). The social sciences, which are focused on "social facts", are epistemologically incapable of considering the "facts of nature" involved in the environmental question (Boudes 2008), which results in a certain conceptual weakness (harmful in the context of interdisciplinarity) and in pragmatic and goal-oriented research. Since the 1990s, this field of research has undergone a significant reversal. While the social sciences were initially criticized for only working on the environment from the perspective of its social construction, it seems that it is now only a natural environment (Kalaora and Vlassopoulos 2013). Today, it is difficult to find a definition of the environment that does not refer to the "natural" environment (Metzger 2017).

Because of the dominant framing by the hard sciences and the limited scope of critical social science approaches, the notion of the Anthropocene emerges from the same trend. In the vision of the world that is imposed, the hard sciences, eminently legalistic, play a fundamental role, despite the recognition of their limits and the growing uncertainties (reflexivity of the *Risk Society*). A scientific category of geology, the term Anthropocene postulates the "human" as a category of the social world.

The "human", an undifferentiated category, refers to a biological condition that denies the social question. The environmental problematic as well as the notion of Anthropocene postulate a community of destiny that corresponds to the scale of the planet, of the history of evolution or of geological time. But as soon as it is a question of concrete situations, of the diversity of the world, the human *we* is an ideology. Stemming from the legitimacy of a biological truth, this idea masks the social differentiations that risk and the environment entail, notably through the contradictory interests and conflicts that they cause. The performative strategy of the human *we* capable of bringing about a form of planetary solidarity between (all) men, (all) societies and the biogeochemical mechanisms of the planet cannot serve as an intellectual agenda for the social sciences. In a world that is more about competition than solidarity, it is however this *we* that imposes itself, producing categories of vision and "division" of the social world that break down the old social categories. These visions, also specific to the *Risk Society*, are anchored in the relationship of men and societies to risks (vulnerable population, humans, women, youth, etc.). This is a way of obliterating the roles of the development of capitalism and globalization (Smith 2010), the violence of North-South relations in the period of colonization and the industrial era (Ferdinand 2019), the responsibility of the neoliberal mode of development in the "degradation of the environment" (Hornborg 2017). In short, it is to evade the systems of exploitation and domination that have shaped the social world.

By producing a generalized discourse on the "human" future and by declining the imperative environmental stakes, from the individual to the planet, the notion of Anthropocene is instituted as a scientific paradigm that relegates to the same limbo the critical knowledge of the social sciences and the emancipatory political action. In this, the autonomy of the social sciences and their ability to formulate the questions of risks and the environment as social and political questions remain crucial. Paradoxically, the more social sciences become involved, the more they are subject to operational objectives and a multidisciplinary framework, the more they tend to adopt a naturalized and managerial approach to the environment (Castree *et al.* 2014).

6.5. Conclusion

Despite its success and the relevance of many points, the *Risk Society* does not lead to a transformation of the world through the advent of a reflexive modernity. Certainly, the risk society is there. Risks, global and local, invisible and materialized, have burst into the public space, around climate change, pesticides or nanotechnologies. All over the world, collective actions of protest or demand on these subjects are flourishing. Multilateral negotiations on environmental issues are punctuating the international political agenda, while speeches and research on the integration of lay knowledge and citizen counter-expertise are multiplying.

But science, technology and innovation remain central. Despite the reflexivity deployed, they continue to be the basis for economic growth as well as for solutions to risks. In the political arena, the democratization of risks is not a current issue. The arenas of negotiation poorly conceal the asymmetry of the actors and the permanence of the hegemony of scientific rationality. In the economic field, the environmental question serves the expansion of capitalism (turning the rights to pollute, or even life itself, into a market).

The *Risk Society* implicitly assumes that decisions are made on the basis of scientific knowledge and rationality, which is at best naive, at worst totally inaccurate. When reflexive modernity speaks of taking seriously the effects of science and technology, or of democratizing risk management, we see the Anthropocene, resilience and sustainable development arriving at the same time, with their share of unavoidable actions.

Faced with an emergency, when “there is no alternative”, the solution that is imposed is not the best one, but the one that remains after we have deprived ourselves of the possibilities of doing otherwise. Understanding the Anthropocene moment presupposes mastering its framework of analysis, and consequently breaking away from the conditions of research production. This insubordination has a cost that the institutional functioning of research makes exorbitant, if not at the margin or far from the action. But who said that emancipation would come from scientific knowledge?

6.6. References

Becerra, S. (2012). Vulnérabilité, risques et environnement : l’itinéraire chaotique d’un paradigme sociologique contemporain. *[VertigO] La revue électronique en sciences de l’environnement*, 12(1) [Online]. Available at: https://journals.openedition.org/vertigo/11988 [Accessed 10 May 2020].

Beck, U. (1992). *Risk Society: Towards a New Modernity*. Sage Publishing, London and New York.

Bensa, A. and Fassin, E. (2002). Les sciences sociales face à l’événement. *Terrain*, 38, 5–20.

Blaikie, P., Cannon, T., Davis, I., Wisner, B. (1994). *At Risk: Natural Hazards, People's Vulnerability and Disasters*. Routledge, London.

Boltanski, L. and Thévenot, L. (1991). *De la justification : les économies de la grandeur*. Gallimard, Paris.

Bonneuil, C. (2002). A plusieurs voix sur La société du risque. *Mouvements*, 21–22(3), 171–174.

Borraz, O. (2013). Pour une sociologie critique des risques. In *Du risque à la menace. Penser la catastrophe*, Bourg, D., Joly, P-B., Kaufmann, A. (eds). PUF, Paris.

Boudes, P. (2008). L'environnement, domaine sociologique. La sociologie française au risque de l'environnement. PhD Thesis, Université Bordeaux 2, Bordeaux.

Boudia, S. and Demortain, D. (2014). La production d'un instrument générique de gouvernement. *Gouvernement et Action Publique*, 3(3), 33–53.

Bouleau, N. (2013). L'excessive mathématisation, symptômes et enjeux. In *Du risque à la menace. Penser la catastrophe*, Bourg, D., Joly, P.-B., Kaufmann, A. (eds). PUF, Paris.

Bourcart, L. (2015). Émergence et usages du concept de résilience dans les mondes académique et instutionnel. PhD Thesis, Université Grenoble Alpes, Grenoble.

Bourdieu, P. (1975). La spécificité du champ scientifique et les conditions sociales du progrès de la raison. *Sociologie et Société*, 7(1), 91–118.

Bourdieu, P. (2001). *Science de la science et réflexivité*. Raisons d'agir, Paris.

Bourg, D., Joly, P.-B., Kaufmann, A. (eds) (2013). *Du risque à la menace. Penser la catastrophe*. PUF, Paris.

Castel, R. (2009). *La montée des incertitudes*. Le Seuil, Paris.

Castree, N., Adams, W.M., Barry, J., Brockington, D., Büscher, B., Corbera, E., Demeritt, D., Duffy, R., Felt, U., Neves, K., Newell, P., Pellizzoni, L., Rigby, K., Robbins, P., Robin, L., Rose, D.B., Ross, A., Schlosberg, D., Sörlin, S., West, P., Whitehead, M., Wynne, B. (2014). Changing the intellectual climate. *Nature Climate Change*, 4, 763–768.

Chauvel, L. and Ramaux, C. (2002). A plusieurs voix sur La société du risque. *Mouvements*, 21–22(3), 166–170.

Coanus, T., Comby, J., Ducheneet, F., Martinais, E. (eds) (2010). *Risques et territoires. Interroger et comprendre la dimension locale de quelques risques contemporains*. Lavoisier, Paris.

Djament-Tran, G. and Reghezza-Zitt, M. (eds) (2012). *Résiliences urbaines, les villes face aux catastrophes*. Éditions Le Manuscrit, Paris.

Dourlens, C., Galand, J.-P., Theys, J., Vidal-Naquet, P. (1991). *Conquête de la sécurité, gestion des risques*. L'Harmattan, Paris.

Dubar, C. (2002). Note de lecture. La société du risque sur la voie d'une autre modernité. *Revue française des affaires sociales*, 2, 211–215.

Ewald, F. (1986). *L'État providence*. Grasset, Paris.

Ewald, F. and Kessler, D. (2000). Les noces du risque et de la politique. *Le Débat*, 2(109), 55–72.

Feenberg, A. and David, J. (2002). Marcuse et l'esthétisation de la technologie. *Quaderni*, 49, 81–101.

Felli, R. (2014). Adaptation et résilience : critique de la nouvelle éthique de la politique environnementale internationale. *Éthique Publique*, 16(1) [Online]. Available at: https://journals.openedition.org/ethiquepublique/1371 [Accessed 14 May 2020].

Ferdinand, M. (2019). *Une écologie décoloniale – Penser l'écologie depuis le monde caribéen*. Le Seuil, Paris.

Gilbert, C. (2003). La fabrique des risques. *Cahiers internationaux de sociologie*, 114(1), 55–72.

Gilbert, C. (2013). Quels risques pour la recherche en sciences humaines et sociales. In *Du risque à la menace. Penser la catastrophe*, Bourg, D., Joly, P.-B., Kaufmann, A. (eds). PUF, Paris.

Gilbert, C. (2016). Revisiter les risques à l'aune de la sécurité ordinaire. In *Faire face aux risques dans les sociétés contemporaines*, Becerra, S., Lalanne, M., Weisbein, J. (eds). Octarès, Toulouse.

Gilbert, C. and Henry, C. (eds) (2009). *Comment se construisent les problèmes de santé publique*. La Découverte, Paris.

Gilbert, C. and Lascoumes, P. (2003). Les politiques des risques en Europe. *Revue internationale de politique comparée*, 10(2), 151–160.

Godard, O. (2004). De la pluralité des ordres – les problèmes d'environnement et de développement durable à la lumière de la théorie de la justification. *Géographie, économie, société*, 3(6), 303–330.

Godard, O., Henry, C., Lagadec, P., Michel-Kerjean, E. (2002). *Traité des nouveaux risques*. Gallimard, Paris.

Henry, C. and Jollivet, M. (1998). Introduction. La question de l'environnement dans les sciences sociales. Éléments pour un bilan. Report, PEVS du CNRS, Paris.

Henry, E., Gilbert, C., Jouzel, J.-N., Marichalar, P. (2015). *Dictionnaire critique de l'expertise. Santé, travail, environnement*. Presses de Sciences Po, Paris.

Hewitt, K. (1997). *Regions of Risk: A Geographical Introduction to Disasters*. Routledge, London.

Hornborg, A. (2017). Dithering while the planet burns: Anthropologist's approaches to the Anthropocene. *Reviews in Anthropology*, 46(2–3), 61–77.

Joly, P.-B. (2001). Les OGM entre la science et le public ? Quatre modèles pour la gouvernance de l'innovation et des risques. *Économie rurale*, 266, 11–29.

Kalaora, B. and Vlassopoulos, C. (2013). *Pour une sociologie de l'environnement : environnement, société et politique*. Champ Vallon, Paris.

Kermisch, C. (2012). Vers une définition multidimensionnelle du risque. *[VertigO] La revue électronique en sciences de l'environnement*, 12(2) [Online]. Available at: https://journals.openedition.org/vertigo/12214 [Accessed 14 May 2020].

Lagadec, P. (1981). *La civilisation du risque. Catastrophes technologiques et responsabilité sociale*. Le Seuil, Paris.

Lahire, B. (ed.) (2004). *À quoi sert la sociologie ?* La Découverte, Paris.

Larrère, C. and Larrère, R. (2013). Peut-on échapper au catastrophisme ? In *Du risque à la menace. Penser la catastrophe*, Bourg, D., Joly, P.-B., Kaufmann, A. (eds). PUF, Paris.

Le Bouter, F. (2014). La sociologie constructiviste du risque de Niklas Luhmann. *Communication et organisation*, 45, 33–48.

Lemoine, B. (2009). Compte rendu : Alain Desrosières, l'argument statistique pour une sociologie historique de la quantification (tome 1) et gouverner par les nombres (tome 2). *Revue d'anthropologie des connaissances*, 3(2), 359–365.

Metzger, P. (2017). Connaissances et relations de pouvoir : les sciences sociales, les risques et l'environnement. HDR thesis, Université Grenoble Alpes, Grenoble.

Metzger, P. and D'Ercole, R. (2011). Les risques en milieu urbain : éléments de réflexion. *EchoGéo*, 18 [Online]. Available at: https://journals.openedition.org/echogeo/12640 [Accessed 31 August 2021].

Metzger, P. and Robert, J. (2015). Environnement et risques : les sciences sociales piégées entre critique radicale et utilité sociale. In *Espaces et rapports de domination*, Clerval, A., Fleury, A., Rebotier, J., Weber, S. (eds). PUR, Rennes.

Murard, N. (2002). A plusieurs voix sur la société du risque. *Mouvements*, 21–22(3), 162–165.

Peretti-Watel, P. (2000). *Sociologie du risque*. Armand Colin, Paris.

Pestre, D. (2004). L'évolution des champs de savoir, interdisciplinarité et noyaux durs. *Natures Sciences Sociétés*, 12, 191–196.

Petit, O., Hubert, B., Theys, J. (2014). Science globale et interdisciplinarité : quand contagion des concepts rime avec confusion. *Natures Sciences Sociétés*, 22(3), 187–188.

Pierret, J. (2008). De la société du risque à la société réflexive. Beck peut-il sauver la modernité ? In *Les ambivalences du risque*, Cartuyvels, Y. (ed.). Presses de l'Université Saint-Louis, Brussels.

Quenault, B. (2013). Retour critique sur la mobilisation du concept de résilience en lien avec l'adaptation des systèmes urbains au changement climatique. *EchoGéo*, 24 [Online]. Available at: https://journals.openedition.org/echogeo/13403 [Accessed 10 May 2020].

Quenault, B. (2015). De Hyōgo à Sendai, la résilience comme impératif d'adaptation aux risques de catastrophe : nouvelle valeur universelle ou gouvernement par la catastrophe ? *Développement durable et territoires*, 6(3) [Online]. Available at: https://journals.openedition.org/developpementdurable/11010 [Accessed 16 May 2020].

Quenault, B., Bertrand, F., Blond, N., Glatron, S., Pigeon, P., Peyrache-Gadeau, V., Rocher, L. (eds) (2011). *Vulnérabilité et résilience au changement climatique en milieu urbain : vers des stratégies de développement urbain durable ?* MSHB, Rennes.

Ravetz, J. (1992). Connaissance utile, ignorance utile ? In *La Terre outragée. Les experts sont formels*, Theys, J., Kalaora, B. (eds). Éditions Autrement, Paris.

Reghezza, M. (2015). De l'avènement du Monde à celui de la planète : le basculement de la société du risque à la société de l'incertitude. HDR thesis, Université Paris 1-Panthéon Sorbonne, Paris.

Revet, S. (2011). Penser et affronter les désastres : un panorama des recherches en sciences sociales et des politiques internationales. *Critique internationale*, 52(3), 157–173.

Revet, S. and Langumier, J. (eds) (2013). *Le gouvernement des catastrophes*. Karthala, Paris.

Rufat, S. (2011). Critique de la résilience pure [Online]. Available at: https://hal.archives-ouvertes.fr/hal-00693162 [Accessed 10 May 2020].

Smith, N. (2010). *Uneven Development: Nature, Capital, and the Production of Space*, 3rd edition. University of Georgia Press, Athens and London.

Theys, J. and Fabiani, J.-L. (1987). *La société vulnérable : évaluer et maîtriser les risques*. Presses de l'École normale supérieure, Paris.

Vandenberghe, F. (2001). Introduction à la sociologie (cosmo)politique du risque d'Ulrich Beck. *Revue du MAUSS*, 17(1), 25–39.

Vibert, S. (2012). Une démocratie sans société ? Critique de Beck et de l'idéologie cosmopolitique en sociologie. *Sociologie et sociétés*, 44(1), 121–140.

PART 3

What Consequences for a Changing Modernity?

7

Understanding the Political Fabric and Effects of Ensemble Flood Forecasts in Europe[1]

Sébastien NOBERT[1], Kristian KRIEGER[2] and Florian PAPPENBERGER[3]

[1]Department of Geography, Université de Montréal, Canada

[2]Institut ISOPOLE, Université catholique de Louvain, Belgium

[3]ECMWF, Reading, United Kingdom

7.1. Introduction

In the wider context of the Anthropocene, in which humans are now seen as a telluric force, images of so-called climate-related disasters such as flooding have become part of daily news and are delivering the promises of spectacle news. In general, by its capacity to unearth the complex relationships between life, science, technology and political institutions, the Anthropocene also exposes tensions in the organization of the relationships connecting humans to non-human processes and elements involved in producing so-called disasters (Clark and Yusoff 2017). With the prospect of a catastrophic changing climate looming ahead, the forecasted changes in the water cycle are predicted to cause an increase in the recurrence and magnitude of floods over the next decades (IPCC 2013). This, in turn, is challenging

1 This chapter partly draws on the work "Understanding the role of modernity, science and risk in shaping flood management", published in 2015 in *Wire Water*, Vol. 2, 245–258. The authors thank Wiley for allowing the content to be reproduced.

Risks and the Anthropocene,
coordinated by Julien REBOTIER. © ISTE Ltd 2021.

the ways in which flood risk is dealt with, questioning the dominance of structural measures such as dykes and flood walls as perhaps less central to adaptive strategies meant to deal with long-term climate-related risks and uncertainties (European Commission 2002). The need to rely less on structural measures is a view shared by international organizations involved in disaster risk management (Pigeon and Rebotier 2016; Vinet *et al.* 2017, see also Chapter 4), such as the United Nations Office for Disaster Risk Reduction (UNDRR), the Emergency Response Coordination Centre of the European Commission (ERCCE) and the European Civil Protection and Humanitarian Aid Operations (ECHO), which are now all promoting the development of instruments such as early warning systems and flood maps as part of climate change adaptation strategies (UNISDR 2006). In fact, the call for further investments in early warning systems to edge against climate- and weather-related uncertainties also surfaces in the last IPCC (Intergovernmental Panel on Climate Change) report on the state of the climate, in which their applications for flood risk management are highly recommended (IPCC 2014). However, understanding how and why the development of instruments such as flood early warning systems became central to adaptation science and policy cannot be understood without unpacking the trajectories leading science and politics to merge in the constitution of anticipatory risk governance, setting the ground for pervasive risk-based approaches.

Thus, by looking at the social and political dimensions of floods, this chapter will show how the modern ideal of predicting the future has favored anticipatory action as a new form of governance within which urgency and insecurity become the main drivers of scientific and political development (see Chapter 8). It is important here to see hydrometeorological predictions as a mode of enquiry that seek to stabilize risk and uncertainties, especially those associated with the Anthropocene. By looking at flood risks, this chapter also argues that the emphasis on predicting hydrometeorological processes has encouraged the development of a quantificational relation to floods that has too often confined us to the logics of cost–benefit analysis and system analysis, which have in turn established mathematics, statistics and behavioral science as the backbones of flood risk governance (Porter 1995; for developments related to urban climate, see Chapter 3). This technocentric vision has prioritized the study of the effects and dangers of flooding rather than challenging its societal causes, leaving us to wonder whether risk is still the right vehicle to engage with the threats posed associated with the Anthropocene moment (Bourg *et al.* 2013), as the challenges posed by the acknowledgement of this new geological era are far more political and philosophical than technical (see Introduction). Our argument unfolds into four interrelated sections.

First, the chapter starts by defining modernity broadly in order to provide a conceptual framework in which our discussion of risk and anticipatory actions is anchored. Second, by looking at the development of numerical weather predictions, the chapter reviews how the dream of predicting future weather became reality through a series of social and political events enabling the consolidation of an ambitious scientific project to simulate the weather. Building on this historical account, the third section focuses on how these transformations affecting the realm of forecasting science have influenced the development of flood forecasting science, particularly the shift from deterministic to probabilistic forecasting with the emergence of the so-called ensemble prediction system (EPS) and their application to hydrology. While reviewing the technical implications of using EPS for flood forecasting, this section engages with some of the consequences of this new science on the institutional dimensions of flood risk management. The fourth section draws on political science to review how the fusion of risk and forecasting science became part of a modern form of governance in the context of flooding, legitimizing risk as a governing practice through anticipatory actions. This demonstration exposes how science and policy meet in the development of probabilistic flood risk instruments meant to inform adaptation to and mitigation of flood risk. Finally, the paper stresses that risk, science and politics have their limitations in helping us to engage with the wider challenges posed by the Anthropocene moment.

7.2. Modernity and the anticipation of risk

In the social sciences, the significance of both modernity and modernization have been widely discussed by sociologists and philosophers in relation to their processes of rationalization (Weber 1992), domestication of nature (Marx 1976) and, more recently, with the work of the late Ulrich Beck (1992), for their effects of individualization. While defining modernity is a rather difficult task, we can broadly define it as a strong division between human society and the non-human world (commonly called nature) in which the former transforms the latter in a spirit of free market, in which individualistic and democratic ideals prevail (Giddens and Pierson 1998), but also, it is a division in which the ancient, archaic and stable are confronted with the new and constantly moving future (Rosa 2013). In modernity, the future is more than a temporal modality; it is the "vanishing point" enabling the rise of a technological rationality that is capable of reshaping collective and individual endeavors (Koselleck 1988). This is illustrated well by the multiple risks and uncertainties that are emerging from the modern era (e.g. climate change). While risks are generally understood as the potential impacts of hazards or processes for which we know the magnitude and the distribution, this chapter sees risks as ways of acting and thinking, having important effects on the shaping of politics and relations with the world.

For Beck (1992), the radicalization of modernity, through an increase in the individualization and secularization of social actors (through scientism), has led to the dissolution of an industrial society to a society in which risk is no longer containable but instead is now diffused and central to the formation of a risk society. This transfiguration of social relations has been made possible by what Beck defines as "reflexive modernization", whereby the consequences of modernization (which refers to the process that lies at the center of modernity, capable of mobilizing the future through the fusion of capitalism, science and scientism) are not captured by modern and democratic political institutions, leading to a plurality of emerging risks and uncertainties, on the one hand, and to a greater public awareness of the secondary effects of techno-scientific developments, on the other hand (see Chapter 6). For Beck (1994), responses to these new forms of uncertainties have turned into a blind faith in technological advances that can now be used as political instruments serving the organization of risk as an underlying principle of modern societies and their futures. The impact of these ideas has been tremendous on the ways in which risk has become an object of study in sociology and political sciences, but also, perhaps more interestingly, in explaining how modernization, through rationalization, has framed particular governance approaches to futures in the context of flood risk management, for example.

In the Anglo-American world, research on risk has been driven mainly by risk analysis, favoring quantitative and metrological approaches ingrained in neoclassical economics and psychology that are associated with cost–benefit analysis, but also with risk perception and acceptability of risk. The proliferation of these approaches is linked to what Rothstein and colleagues (2006) have defined as the "risk colonization" process, whereby risk becomes an important category used in reshaping politics and public policies in almost all aspects of governance, ranging from terrorism to flood risk. Indeed, the shift toward a dominant culture of risk is also the product of a co-production of politics and science through which experts who were in transition between academia, government and think tanks were advocating for the application of rational choice theory in all sectors of the state (Green and Shapiro 1994; Boudia 2013). This rationalization has led to policymaking being instrumentalized through a battery of quantitative tools and quantificational practices that have made risk more than a simple matrix, but has become a way to govern and relate to life more generally.

Investigating risk as a form of governance implies that we should also look at the modality that enables modern societies to prepare for dangerous events through anticipatory actions. Framed in the modern ideal, anticipatory actions are primarily animated by the desire to protect liberal-democratic life from threats of all sorts, such as flooding. This is done by assembling different tools and techniques, which above all incarnates the modern ideal of grasping the future before it happens, and thus controlling its course and its nature in order to prevent disasters (Anderson 2010). Thus, anticipation is a core quality of modern liberal governance regimes, which in turn are exposing social actors to uncertainties and indeterminacies in the state of knowledge that are meant to define and organize future risks.

In order to understand how anticipation becomes foundational to current flood risk management, there is a need to better identify how the vision of the future as a graspable and controllable entity becomes possible and key to our relationship with flooding. The following section will look back at the beginning of computation and modern scientific advances allowing the prediction and anticipation of weather through numerical weather predictions (NWP). We will then show how those developments are taking place in flood forecasting and policy.

7.3. Numerical weather predictions and the emergence of forecasting the future

In 1922, the British mathematician Lewis Fry Richardson published *Weather Predictions by Numerical Process*, a book that became foundational to numerical forecasting and to the modern quest of predicting the future (Lynch 2006). By forcing arithmetical procedure to solve differential primitive equations developed by meteorologists to describe the physics of the atmosphere, Richardson pushed forward the computation of mathematical methods meant to understand atmospheric motions in time. This mathematical intrusion allowed Richardson to move from forecast operations based on variables to operations based on numbers, enabling him to propose an algorithm (i.e. a group of operations carried out sequentially) capable of creating forecasts based on numerical data and to provide an answer to what the father of modern meteorology, Vilhelm Bjerknes, had defined earlier as "the one problem worth attacking… the precalculation of future conditions" (Lynch 2000, p. 4). It is also in this book that Richardson famously stated that

> perhaps some day in the dim future it will be possible to advance computations faster than the weather advances and at a cost less than

> the saving of mankind due to the information gained. But this is a dream. (Richardson 1922, p. vii)

Although the book did not create the enthusiasm desired by Richardson at the time of publication, and was criticized for the significant errors in the basic equations of atmospheric hydrodynamics, Richardson's contribution to formulating the process of numerical forecasting has made both his dream and his ambitions vital to the development of meteorology and forecasting more generally.

As a result of military demand during the Second World War to improve knowledge of weather systems, and the dreams of controlling the weather that lay at the heart of the Cold War, the subsequent deployment of an arsenal of financial, technical and intellectual resources dedicated to the improvement of weather forecasts have radically changed the face of weather forecasting during the 20th century (Goldstine 1980; Harper 2008). Those changes have not only led to the unification of meteorology by grounding its principles in physical theory and thus reshaping the discipline into a "formal science", but they also opened up the field to new mathematics interested in the numerical computational methods used to solve hydrodynamics and aerodynamics problems, most of them inherited from ballistic and thermonuclear research (Aspray 1990). Along with the development of high-speed electronic computers capable of implementing primitive equation models, the Cold War period was vital to the improvement of forecasts' algorithms linked to the refinement of global circulation modeling (see also Chapter 5). These constant advancements of numerical techniques transfigured meteorology between the 1950s and the 1970s in two significant ways. First, by making numerical weather forecasting central to research programs in meteorology, a set of new institutions and networks needed to be developed in order to establish a vast campaign of weather data collection that became necessary for forecasting, and this had the underlying effect of reshaping the institutional organization of weather forecasting around the world (Edwards 2010). For example, the European Centre for Medium-Range Weather Forecasts (ECMWF)[2] was formed in 1975 to address the challenge of medium-range (3–15 days) forecasting and became a world leader in producing and disseminating numerical weather forecasts worldwide. Second, by building heavily on complex theories of physics and mathematics to describe the non-linearity of atmospheric processes, numerical weather forecasting needed to benefit from the increasing power offered by refined electronic computing. This technical contribution to the development of numerical predictions has made the

2 See https://www.ecmwf.int/en/about for additional information about the center, its aims and institutional setting [Accessed 16 June 2020].

pillars of numerical forecasts, such as the mathematician John von Neumann and the meteorologist Jules Charney, believe that electronic computers could become new experimental and inductive devices for both operational and research meteorology (Aspray 1990; Nebeker 1995). These two changes have converged to make the dream of predicting the future of atmospheric processes possible through a rapid series of theoretical innovations, but they have also contributed to making computer science fundamental to the development of meteorology in the decades that followed. This latter implication is far from being trivial. On the contrary, it has also engendered wider transformations in the scientific community more generally, demonstrating the potential to run various types of mathematical models through a complex set of algorithms on high-speed computers.

Finally, another giant step in the story of numerical weather predictions comes from the work of meteorologist Edward Lorenz (1993), who pointed out that the atmosphere is a nonlinear chaotic system in which the initial state is unknown. Following Lorenz's work, it is now understood that slight changes in the initial conditions can lead to significant differences in forecasting results. Building on the increasing computer power and subsequent development of numerical weather predictions, Lorenz made it possible to bring a new set of scientific inquiries to the center of meteorological questions on the validity of physical models' equations and the errors of individual prediction models (Thompson 1984, p. 1). In practice, this meant that rather than issuing a single forecast, multiple forecasts were needed to represent future states and cover the probabilities caused by the uncertainties. Multiple forecasts required an increase in computational power. The combination of these new possibilities provided by chaos theory and the continually increasing computer power allowed meteorologists to include, and thus model, key factors which had previously limited the predictability of the weather, including uncertainty in the initial conditions or in the numerical presentation of physical processes. This led to the development of new types of probabilistic forecasting (Molteni *et al.* 1996; Buizza *et al.* 1999).

While the emergence of numerical predictions exposed some fragments of the long and complex history of forecasting in meteorology, the development and application of numerical weather predictions, along with computer power and the need for weather data, contributed to the fusion of climatology and meteorology during the 20th century. However, as we are going to discuss in the following section, this fusion has extended to other fields; hence, knowing how numerical weather predictions have developed is of cardinal importance to fully understand their impacts on the development of flood forecasting techniques in the 1990s and their integration into wider risk management strategies.

7.4. Flood forecasting, ensemble predictions and probabilistic risk management

Numerical weather predictions have also played a part in the emergence of present-day flood forecasting systems, which are now increasingly moving toward integrating medium-term (4–15 days ahead) probabilistic meteorological forecasts (Buizza *et al.* 1999). If high-speed computers and simulations are now becoming experimental sites that are used to understand flood behavior and simulate their potential extent and moment of occurrence, the universe of flood forecasting is still sharply divided into two different schools of thoughts or traditions: (i) deterministic forecasting and (ii) probabilistic forecasting (Figure 7.1). However, in this chapter, we focus solely on probabilistic forecasting.

Knowing how to cope with uncertainty in flood forecasting has always been seen as an important source of concern on which hydrologists have concentrated their efforts in the development of methods such as the Ensemble Streamflow Prediction (ESP) system put in place in the 1970s. This allowed them to use deterministic hydrological and scenario-based models that used historical data, which in turn made it possible to generate an ensemble of streamflow hydrographs (Day 1985). While the initial idea did not use meteorological forecasts, ESP forecasts were used to compare the current and recent past simulations and observations. They assessed how similar the current hydrological situation was to historic situations and then picked a number of those historic situations as a forecast for the future, fusing uncertainties from both meteorology and hydrology together. While these systems are widely used today for seasonal weather forecasts (e.g. in Australia and the United States), both medium- and short-range flood forecasting systems remained stubbornly deterministic (Krzysztofowicz 2001). Although the development of probabilistic forecasts has generated significant debates among hydrologists since the mid-1990s (Cloke and Pappenberger 2009), the number of probabilistic flood forecasting systems currently operational in the world is still very low in spite of a net popularity for their development, which can be attributed, on the one hand, to the general acceptance in the flood forecasting community that forecasts always contain uncertainties. On the other hand, this acceptance of uncertainty among the hydrological community is also the product of the wider "risk colonization" process that is forming through the development of mathematics meant to improve risk analysis and decision making under uncertainty, such as Bayesian statistics, which enable the production of scenarios with different parameters.

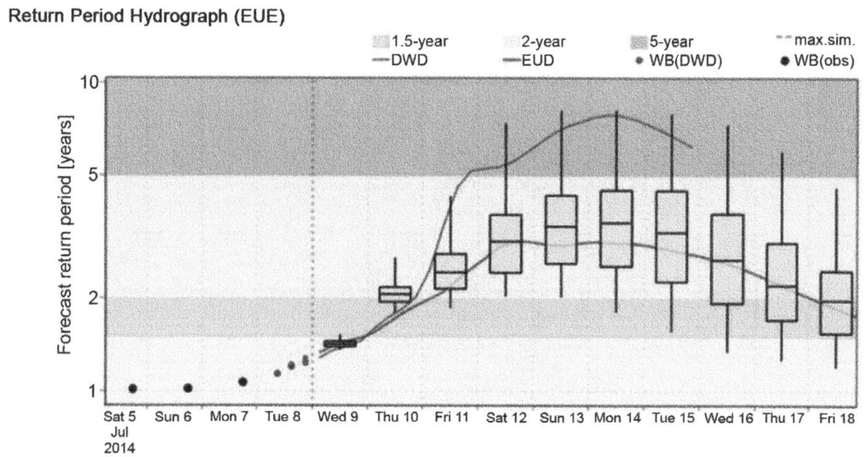

Figure 7.1. *Deterministic and probabilistic forecasts from the European Flood Awareness System. For a color version of this figure, see www.iste.co.uk/rebotier/risks.zip*

The blue solid line is a flood forecast using the deterministic model of the German weather service (Deutscher Wetterdienst). The red solid line is the high-resolution forecast of ECMWF and the box-plots show the ensemble forecast of ECMWF.

Additionally, advances in meteorological sciences resulting from the rapid development in numerical weather predictions and the unceasing increase in computer power opened the door to new forecasting techniques, such as the development of the EPS at the beginning of the 1990s (Tracton and Kalnay 1993), whose main characteristic is to provide multiple forecasts for the same time period (Figure 7.2). EPS forecasts are known for their added value in terms of greater skill and longer lead time, but also for their capacity to generate a suite of forecasts that provide a way to quantify forecasts' uncertainties (Palmer 2002; Demeritt *et al.* 2010). By using EPS, it is now possible to account for the uncertainty of numerical weather predictions that are used in hydrological flood forecasting, a possibility that did not exist before the beginning of the 2000s. Currently, there are multiple meteorological centers across the globe issuing EPS, most of them generating uncertainties based on different methods (e.g. models, data assimilation, ensemble generation, etc.), which are in turn used to improve hydro-meteorological flood forecasting through a combination of techniques.

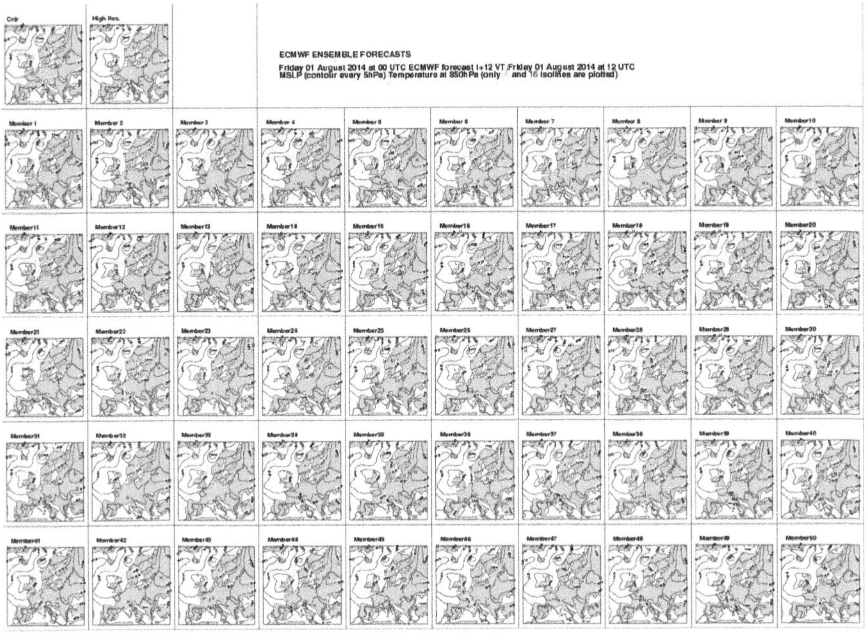

Figure 7.2. *Simultaneous ECMWF forecasts. For a color version of this figure, see www.iste.co.uk/rebotier/risks.zip*

The figure shows 50 perturbed ensemble members, the high-resolution operational model and the control run that can be seen in the above two top left images. The control run is similar to the high-resolution run as it is produced without perturbations, but with the same resolution than individual members, which are produced at a lower resolution than the high-resolution perturbations images. Those differential runs are created to provide a reference run making it possible to compare differences in forecasts depending on resolution. When talking about the full ensemble, this usually means an inclusion of the control run, making it 50 + 1 members or 51 members.

EPS and their transformation into ensemble flood forecasts have generated new constellations of networks involved in refashioning the institutional organization of flood forecasting and the meteorological information needed in the management of extreme weather. This can be seen through a myriad of new channels through which those forecasts are now produced, disseminated and communicated. A good example of this new organization of flood information is the European Flood Awareness System (EFAS) (Figure 7.3), which has been developed – and politically defined –

as a technological response by the European Commission to the 2002 floods of the Danube, Elbe and Oder rivers, which were responsible for displacing 500,000 people and killing more than 80 (Demeritt *et al.* 2010). Although EFAS was created and made operational by the European Joint Research Centre, the EPS used to generate its forecasts were produced by the European Centre of Medium-Range Weather Forecasts (ECMWF), which were then used to disseminate innovative ensemble flood forecasts to the 29 forecasting agencies that have signed the Memorandum of Understanding (Demeritt *et al.* 2013).

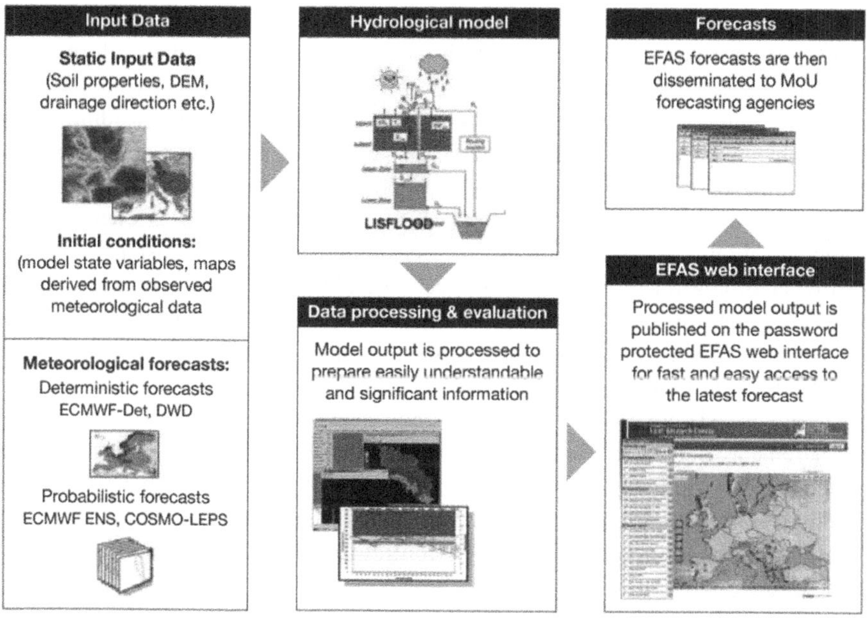

Figure 7.3. *Schematic view of the EFAS system*

There are several hydro-meteorological centers across the globe that are now issuing ensemble flood forecasts; although there were only a handful of centers at the beginning of the century (Cloke and Pappenberger 2009), there were over 20 in 2014, and they are changing the geography of scientific production and the dissemination of flood forecasting. Since generating EPS demands computer power that is capable of performing additional post-processing in order to remove systematic biases, most centers interested in developing ensemble flood forecasts need to merge other data, models and forecasts with their own "in house" models and products, resulting in what Edwards (2010; Edwards *et al.* 2011) has termed "data friction". These "data frictions" correspond to further tensions about the ways in which the coarse scale of meteorological EPS should be downscaled statistically into a hydrological model and, when this is done, to how those new forecasts are then used in relation to "in house" models. These "frictions" have been identified as one of the factors limiting predictability (other factors include catchment characteristics, e.g. catchment response time) and impacting on these novel ensemble flood forecasting systems. Thus, behind the promises of better forecasting skills, ensemble flood forecasts are slowly reshaping the world of flood forecasting, implying a shift from a risk culture based on deterministic thinking and risk prevention to one based on probabilistic knowledge and preparedness for risk, and hence enforcing a new kind of organization of risk and management strategies yet to come (Demerit and Nobert 2011). With the possibility of providing informational value on uncertainties and the capacity to predict future extreme weather through enhanced numerical weather predictions, the scientific trajectory leading to the development of ensemble flood forecasts becomes an example of how the colonization of quantifiable risk has been changing the practice of hydrological forecasting and management of flood events. For example, it is clear that the communication of probabilities is changing the relationships between hydrometeorological services, civil protection and the so-called non-experts meant to receive this information, while it is also producing new sets of institutional and professional challenges that are still relying on deterministic information and yes and no decision making (Demeritt and Nobert 2011). The next section will expose how risk-based governance has not only accelerated the quantification and measurement of dangers, but has also situated insecurities and dangers as focal relations to the Anthropocene moment, raising important concerns about whether its ontological essence of risk is fit to face the current indeterminacy that underlies this "new" era as we are more likely to be confronted to recurrent and extreme flood events.

7.5. The promises and pitfalls of risk-based flood risk management

While the development of risk as an instrument of governance can be tracked down to the use of probabilities by the insurance sector from the 17th to the 19th centuries (Ewald 1986; Porter 1995; Desrosières 1998), it is not before the 1970s and 1980s that its predominance in shaping new categories affecting both public and private spheres of governance could really be felt in Western liberal democracies. With neoconservative regimes flying high during the mid-1980s in the Anglo-American world (Reagan in the United States and Thatcher in the UK), risk was not only a discourse extending the economy to every dimension of social life, but also a way of being. In many advanced capitalist societies, but more particularly in Anglo-American contexts, it has been argued that risk- and evidence-based approaches to governance are on the rise (Rothstein *et al.* 2006; Boudia and Jas 2007). For instance, under the New Labour government in Britain, risk was to become "*the* governing concept"[3] in all changes of regulation (Power 2004). Beyond Britain, the EU has been active in promoting risk-based approaches to governance. The European Commission, for example, notes that

> risk governance – embracing risk identification, assessment, management and communication – has become a crucial [...] component of public policy. (European Commission 2002, p. 23)

In Europe, the use of risk as a governance concept extended to flood-related disaster management in the 1990s. This was largely related to the increasing number of flood events occurring during this decade, and with the subsequent record-breaking of economic damages and loss of lives, which in turn triggered a battery of engineering-driven solutions (e.g. dykes, flood walls, dredging) to deal with flood risk, which became challenged by new forms of technological and political responses, particularly flood warnings and preparedness strategies. This progressive change of mentality became more perceptible in many European countries where governments sought this transformation through programs such as "learn to live with floods" (ICE 2001), "make space for water" (in the UK) (DEFRA 2005), "room for rivers" (in the Netherlands) and "space for rivers" (in Germany) (Berendes 2005) at the beginning of the 21st century (Johnson *et al.* 2005). More concretely, these technocratic strategies were driven by economic theory and sought to adapt socioeconomic processes and progress to biophysical processes, through a combination of long-term (e.g. land-use regulation in flood plains) and short-term (e.g. flood forecasting) interventions. In conjunction with scientific developments meant to reinforce the calculability of risk (such as EPS), these long- and short-term strategies aimed to deliver adaptation to climate change while protecting populations from floods at a lower cost than structural measures. However, it is also important to see the trajectory of sophisticated risk instruments such as EPS as well as a risk-based approach to governance as powerful agents reforming the institutional geographies of flood risk management, which in the same trend are allowing specific sets of values at the heart of the European project becoming effective in practice such as the use of probabilistic forecasting in flood management. This is not only an innovative risk management approach, but it is also allowing the EU through the European Commission to maintain its image of scientific leaders.

Risk-based approaches can be simply summarized as being a blend of governance resources, such as information, regulation and authority, and financial

3 Tony Blair's speech on compensation culture, 26 May 2005 [Online]. Available at: http://www.guardian.co.uk/politics/2005/may/26/speeches.media [Accessed 13 June 2020].

resources which are allocated on the basis of risk levels that are previously identified by scientists and later on used by regulators or policymakers seeking better organizational control. A second characteristic of these risk-based approaches is the recognition of the uncertainty that surrounds risk assessments, as we saw in relation to the identification of uncertainties in flood forecasting through EPS4. While the development of flood forecasting technologies and flood hazard maps are only one element of the adoption of risk-based approaches to flood management pushed by the EU, these approaches are instrumental for planning emergency response measures (e.g. evacuation procedures, preparing pumps, having enough staff on standby, etc.) or increasing the preparedness of those members of the public judged as being at risk. Benefitting from the same scientific advances and models as flood forecasting, flood hazard maps displaying spatial zones that might be inundated in the future are showing a different probability of occurrence and associated material damage enable governments, for instance, to regulate land use and construction in accordance with different levels of risk (Figure 7.4). Flood mapping has become an integral part of the recent flood-related legislation in Europe. For example, in the EU Floods Directive (European Commission 2007), EU member states were required to produce flood risk maps, undertake risk assessments and prepare risk management plans so that they were ready for 2015.

In addition to anticipating and managing specific threats, a risk-based approach uses risk as a tool for optimizing the coordination, prioritization and effectiveness of governance. Economic calculus is still central here, so that the effectiveness and efficiency of public interventions are ensured through (risk-)proportionate allocation of resources. For example, in a strategy document on environmental management for the "21st century" (Environmental Agency 2005), the UK's Environment Agency (with mandates limited to England and Wales) highlights that resources allocation should be better targeted and proportionate to interventions and that this should maximize benefits while reducing the regulatory burden. Similar arguments are made by the Organisation for Economic Cooperation and Development (OECD) in which risk-based governance implies being capable of "targeting mitigation investments to their greatest benefits" (OECD 2009, p. 5).

4 It is important to keep in mind that by their capacity to recognize the uncertainties related to risk assessments and regulations, risk-based approaches are involved in shaping what Michael Power (2004, p. 89) has defined as "the politics of uncertainty", whereby the possibility for failures and the imperfections of regulations and policies are accepted and made public. Risk-based approaches are common in the UK while they are less common in countries such as France and Italy where failures to protect lives have led to criminal prosecutions. For example, we can cite the well-publicized cases of the 2009 Xynthia storm and related floods in France and the 2009 L'Aquila earthquake in Italy for which scientists and civil servants were trialed for failing to prevent the deaths of citizens.

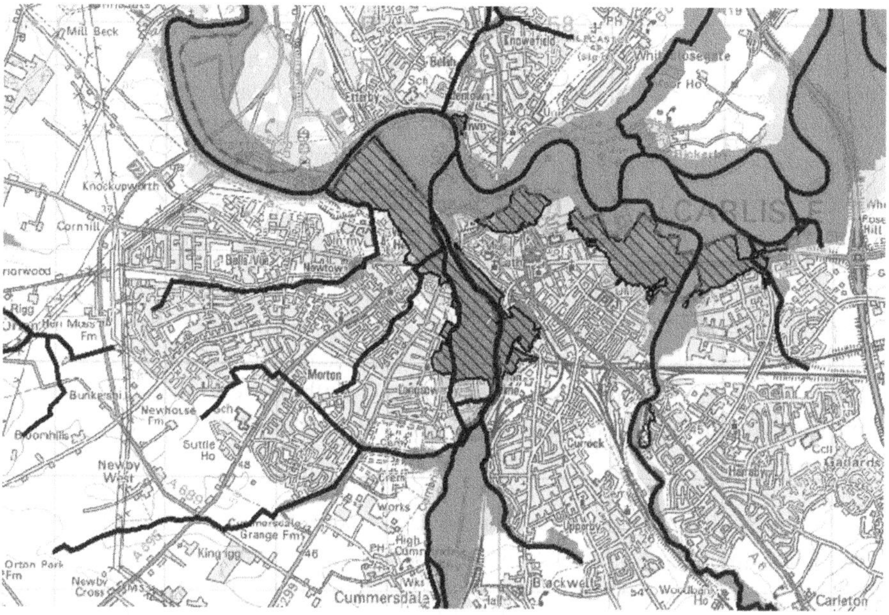

Figure 7.4. *Example of flood risk mapping taken from the river Eden in the UK. For a color version of this figure, see www.iste.co.uk/reboulet/risks.zip*

These optimization strategies suggest that there is more to the rise of risk-based governance than simply the availability of technology, scientific advances and the state's interest in managing threats, such as flooding, more effectively. In fact, the interaction between politics and science creates new forms of governance, in which technological means are promoted and used through a rationalization process meant to transform public policies and to shape collective futures. More precisely, risk-based approaches to governance can be associated with three overlapping types of rationalization processes: (i) rationalization through science, (ii) rationalization through economics and (iii) rationalization through accountable governance. However, in the context of the Anthropocene, all these rationalization processes raise questions about the limits of risk-based governance and risk instruments that deserve to be highlighted.

With its roots in positivism and quantification methods, the first rationalization process aims at building trust in and objectivity around modern institutions while maintaining ontological security. The underlying idea is that science-based and/or evidence-based decision making is a superior and objective way of solving the problems of modern society. In the Anthropocene moment, this first type of rationalization process uses scientific disciplines such as hydrology and hydraulic engineering to provide crucial inputs to understanding the properties of flood hazards and how to deal with them, while ignoring the non-static dynamics that shape this novel era and for which there are no historical data to rely on to predict its course, hence limiting the efficiency for this rationalization process in evaluating and predicting the emergence of new and more complex types of risks. While this first rationalization process allows us to see risk as a category of action, or a particular lens through which a situation is being analyzed and framed by drawing on mathematics and statistical methods, the practical limits to this rationalization process remain problematic. For example, the use of flood risk assessments supported by mathematical models are seeking to identify behavioral and regularized patterns in flood events while abrupt changes in those patterns are what defines the hydrometeorological effects of the Anthropocene moment, therefore impairing the predictive and anticipatory abilities assumed by this rationalization process.

Also grounded in mathematics and disseminated by economists working on risk, economic rationality is associated with neoliberal "governmentality" or, put simply, with the ensemble of practices whereby individual freedoms, rights and relationships to others can be defined, organized and instrumentalized by a form of governance that has neoliberalism as its main engine (Foucault 1994). The most common expression of this kind of rationality is the so-called cost–benefit analysis, which provides quantitative, comparable, often monetary values of the adverse outcomes of different courses of action. In the context of flood management, such rationality takes the shape of investments in the development of structural measures such as dykes and, more recently, new flood forecasting systems allowing both the protection of private properties and the evacuation of people as well as the protection of goods before the flood peak. It is also a form of rationality that has played a significant role in the colonization of risk to several areas, including flood management. However, in the highly friable grounds produced by the Anthropocene, this kind of rationality is assuming the continuity of neo-liberal "normality" and the necessity to develop metrological risk management as social practice. The political challenges brought by the Anthropocene moment call for a qualitative valuation of many problems including ethical choices to be made on what and who should be saved or be protected before, during and after flood events, displacing the quantitative morals and values that are dominating risk-based governance and replacing them with different ways of assessing and understanding a world on a path of irremediable changes. It is therefore becoming difficult to put a price on the value of life and of some biophysical relations that made our world, but that needs to be considered in a decision making that can perform well within an era defined by its uncertainties.

The third rationalization argument is concerned with the way in which risk can contribute to a greater legitimacy of governance by making the limits of governance more explicit (Rothstein *et al.* 2013; Krieger 2013). Set in a context of high accountability and transparency pressures of modern societies, as well as rampant blame games in the case of risk management failures, policymakers face significant institutional and political challenges when managing risks such as flooding. Those challenges can range from declining popularity and judicial proceedings to budget cuts and trimming the responsibilities of agencies to fit with individual resignations. In the case of flooding, the transfer of responsibility from the state to the individual by the use of probabilistic forecasting is transforming power relationships whereby risk becomes used by non-experts to challenge their position on flood risk and to reinstate the category of risk as the main realm within which floods are dealt with. While probabilities of risk cannot be reduced to a simple yes or no answer meant to ease decision making, risk-based governance assumes that by transferring collective and governmental responsibilities to the individuals, decision making will be better adapted to differential needs in addition to provide more honest information and scenarios. These technocratic ambitions assume that the modern value of individuality will remain key in the face of great shocks that have the capacity to reinvent the political such as the complex side effects of the Anthropocene. This assumption is in turn undermining how collective actions and responsibilities remain often essential in the face of adversities and great uncertainties as the world has witnessed with COVID-19, through which governments were called to help more than ever while a sense of community became indispensable to deal with isolation during lockdown periods.

Finally, while risk instruments enable the ideal of modernity to take shape through different processes of economic quantification, optimization and rationalization, the development, refinement and application of technological means such as probabilistic flood forecasting methods or flood hazard mapping are becoming powerful ways to organize collective relationships to floods. However, it is important to note that beyond the tensions we have illustrated above, risk-based approaches to governance have been widely contested. One important argument comes from the sociology of organizations and builds on the criticism that risk-based approaches to governance are incompatible with institutionalized values, structures and procedures that take place in different national contexts (Douglas and Wildawsky 1983; Jasanoff 2005). Thus, different attitudes to risks and governance can come to undermine any attempt to provide a universal model of a risk-based approach to flooding. Notions such as trust in the state's ability to provide security to its population, the role of the insurance industry and the significance of private property can vary enormously from one state to another, leading to different responses to risk (Pottier *et al.* 2005). More recently, the Francophone sociology of risk has also questioned the validity of a risk society as a concept relevant to the issue of contemporary threats (Bourg *et al.* 2013), arguing that recent historical enquiries in research relating to the development of risk have demonstrated that risk society and environmental reflexivity are not the products of contemporary societies and postmodernity, but, rather, phenomena which have persisted since the 18th and 19th centuries (Locher and Fressoz 2012).

7.6. Flood risk management in the Anthropocene moment

By stepping into the moment that defines the Anthropocene, not only have humans entered a period of unprecedented changes, but they are also moving into an epoch in which the limits to science and evidence-based policy are becoming obvious. The major uncertainties and indeterminacies emerging from the consequences of global temperature changes, biodiversity loss and extreme weather events such as those exposed through the example of flood risk management have made it possible to underline the limits of risk as a concept of governance while revealing the importance of modernity as a conceptual matrix capable of explaining the evolution of flood risk management. By looking at the development of both predictive science and flood risk governance regimes, the chapter shows how the idea/dream of foreseeing the future scientifically and controlling its outcomes politically becomes possible. More interestingly perhaps, for the moment that defines the Anthropocene, the analytical exercise provided by the chapter shows how technological optimism as well as risk-based governance are building on the old foundations of so-called normality that are now less relevant for thinking about political aspects in the indeterminate era that defines the one we currently live in.

We have demonstrated that the early scientific dream of predicting the future was animated by the belief in modern capacities to stabilize and organize our relation to the future and uncertainties in the state of knowledge, allowing us to speed up the now through anticipative actions tailored for a dangerous and unstable not-yet such as the one defining the Anthropocene moment and its finality. We have also shown that the spectacular scientific development in computation was aided not only by scientific ingenuity but also by the particular contexts within which military interest in mastering the weather has played a significant role. The spillover of forecasting from meteorology and climatology to flood management demonstrates how increasingly institutionalized scientific methods and approaches have helped to frame risk as an instrument of governance that is capable of impacting various scientific and technological fields while changing how scientists and practitioners are approaching and understanding the world. Yet, the general shift toward anticipatory forms of governance based on risk illustrates also the problem of reinventing our ways of knowing and questioning anew in a moment requiring radical change in a way we experience and act upon our world. The strength of risk-based governance is also its capacity to create a "quantificational locked-in syndrome", whereby the challenges to rethink the ways in which we relate to each other in the Anthropocene require re-evaluation of old sets of morals and values and questioning what is important outside numbers, spreadsheets, predictive instruments and economic normativity.

7.7. References

Anderson, B. (2010). Preemption, precaution, preparedness: Anticipatory action and future geographies. *Progress in Human Geography*, 34(6), 777–779.

Aspray, W. (1990). *John von Neumann and the Origins of Modern Computing*. MIT Press, Cambridge, MA.

Beck, U. (1992). *Risk Society: Towards a New Modernity*. Sage, New Delhi.

Beck, U. (1994). The reinvention of politics: Towards a theory of reflexive modernization. In *Reflexive Modernization: Politics, Tradition and Aesthetics in the Modern Social Order*, Beck, U., Giddens, A., Lash, S. (eds). Polity Press, Cambridge.

Berendes, K. (2005). Das Hochwasserschutzgesetz des Bundes. *Zeitschrift für Wasserrecht*, 44/45, 197–214.

Boudia, S. (2013). La genèse d'un gouvernement par le risque. In *Du risque à la menace*, Bourg, D., Joly, P.B., Kaufmann, A. (eds). PUF, Paris.

Boudia, S. and Jas, N. (2007). Risk and risk society in historical perspective. *History and Technology*, 23(4), 317–331.

Bourg, D., Joly, P.B., Kaufmann, A. (eds) (2013). *Du risque à la menace*. PUF, Paris.

Buizza, R., Miller, M., Palmer, T.N. (1999). Stochastic representation of model uncertainties in the ECMWF ensemble prediction system. *Quarterly Journal of Royal Meteorological Society*, 125, 2887–2908.

Clark, N. and Yusoff, K. (2017). Geososcial formations and the Anthropocene. *Theory, Culture & Society*, 34(2–3), 3–23.

Cloke, H.L. and Pappenberger, F. (2009). Ensemble flood forecasting: A review. *Journal of Hydrology*, 375, 613–626.

Day, J. (1985). Extended streamflow forecasting using NWSFRS. *Journal of Water Resources and Management*, 111(2), 157–170.

DEFRA (2005). Making space for water. DEFRA, London [Online]. Available at: https://webarchive.nationalarchives.gov.uk/20130403030709 and http://archive.defra.gov.uk/environment/flooding/documents/policy/strategy/strategy-response1.pdf [Accessed 13 June 2020]

Demeritt, D. and Nobert, S. (2011). Responding to early flood warning in the European Union. In *Forecasting, Warning, and Transnational Risks: Is Prevention Possible?* Meyer C.O., De Franco, C. (eds). Palgrave, London.

Demeritt, D., Nobert, S., Cloke, H., Pappenberger, F. (2010). Challenges in communicating and using ensembles in operational flood forecasting, *Meteorological Applications*, 17(2), 209–222.

Demeritt, D., Nobert, S., Cloke, H., Pappenberger, F. (2013). The European Flood Alert System and the communication, perception, and use of ensemble predictions for operational flood risk management. *Hydrological Processes*, 27(1), 147–157.

Desrosières, A. (1998). *The Politics of Large Numbers: A History of Statistical Reasoning.* Harvard University Press, Cambridge, MA.

Douglas, M. and Wildawsky, A. (1983). *Risk and Culture.* University of California Press, London.

European Commission (2002). Regulation N°178 laying down the general principles and requirements of food law, establishing the European Food Safety Authority and laying down procedures in matters of food safety [Online]. Available at: http://eurlex.europa.eu/legal-content/EN/ALL/?uri=CELEX:32002R0178 [Accessed 13 June 2020].

European Commission (2007). Directive 60 on the assessment and management of flood risks [Online]. Available at: http://eur-lex.europa.eu/legal-content/EN/TXT/?uri=CELEX:32007L0060 [Accessed 13 June 2020].

Edwards, P.N. (2010). *A Vast Machine: Computer Models, Climate Data, and the Politics of Global Warming.* MIT Press, Cambridge, MA.

Edwards, P.N., Mayernik, M.S., Batcheller, A.L., Bowker, G.C., Borgman, C.L. (2011). Science friction: Data, metadata, and collaboration. *Social Studies of Science*, 41(5), 667–690.

Environmental Agency (2005). Delivering for the environment: A 21st century approach. *Environment Agency*, Bristol [Online]. Available at: http://www.oecd.org/environment/outreach/33947795.pdf [Accessed 13 June 2020].

Ewald, F. (1986). *L'État Providence.* Grasset, Paris.

Foucault, M. (1994). L'éthique du souci de soi comme pratique de la liberté. In *Michel Foucault Dits et écrits II : 1976–1988*, Defret, D., Ewald, F., Lagrange, F., (eds). Gallimard, Paris.

Giddens, A. and Pierson, C. (1998). *Conversations with Anthony Giddens: Making Sense of Modernity.* Polity Press, Cambridge.

Goldstine, H.H. (1980). *The Computer: From Pascal to von Neumann.* Princeton University Press, Princeton, NJ.

Green, D.P. and Shapiro, I. (1994). *Pathologies of Rational Choice Theory: A Critique of Application in Political Science.* Yale University Press, New Haven, CT.

Harper, K. (2008). *Weather by the Numbers: The Genesis of Modern Meteorology*. MIT Press, Cambridge, MA.

ICE (2001). *Learning to Live with Rivers*. ICE, London.

IPCC (2013). *The Physical Science Basis. Contribution of Working Group I to the Fifth Assessment Report of the Intergovernmental Panel on Climate Change*, Stocker, T.F., Qin, D., Plattner, G.-K., Tignor, M., Allen, S.K., Boschung, J., Nauels, A., Xia, Y., Bex, V., Midgley, P.M. (eds). Cambridge University Press, Cambridge.

IPCC (2014). *Impacts, Adaptation, and Vulnerability. Part A: Global and Sectoral Aspects. Contribution of Working Group II to the Fifth Assessment Report of the Intergovernmental Panel on Climate Change*, Field, C.B., Barros, V.R., Dokken, D.J., Mach, K.J., Mastrandrea, M.D., Bilir, T.E., Chatterjee, M., Ebi, K.L., Estrada, Y.O., Genova, R.C., Girma, B., Kissel, E.S. Levy, A.N., MacCracken, S., Mastrandrea, P.R., White, L.L. (eds). Cambridge University Press, Cambridge.

Jasanoff, S. (2005). *Designs of Nature–Science and Democracy in Europe and the United States*. Princeton University Press, Princeton, NJ.

Johnson, C.L., Tunstall, S.M., Penning-Rowsell, E.C. (2005). Floods as catalysts for policy change: Historical lessons from England and Wales. *International Journal of Water Resources Development*, 21(4), 561–575.

Koselleck, R. (1988). *Critique and Crisis: Enlightenment and the Pathogenesis of Modern Society*. Berg, Oxford.

Krieger, K. (2013). The limits and variety of risk-based governance: The case of flood management in Germany and England. *Regulation & Governance*, 7(2), 236–257.

Krzysztofowicz, R. (2001). The case for probabilistic forecasting in hydrology. *Journal of Hydrology*, 249(1–4), 2–9.

Locher, F. and Fressoz, J-B. (2012). Modernity's frail climate: A climate history of environmental reflexivity. *Critical Inquiry*, 38(3), 579–598.

Lorenz, E.N. (1993). *The Essence of Chaos*. University of Washington Press, Seattle, WA.

Lynch, P. (2000). Numerical weather prediction: The origins and development of computer forecasting. Historical note No. 4. *Met Éireann*, August 4–7.

Lynch, P. (2006). *The Emergence of Numerical Weather Prediction: Richardson's Dream*. Cambridge University Press, Cambridge.

Marx, K. (1976). *Capital*, Volume 1. Penguin, London.

Molteni, F., Buizza, R., Palmer, T.N., Petroliagis, T. (1996). The new ECMWF ensemble prediction system: Methodology and validation. *Quarterly Journal of Royal Meteorological Society*, 122(529), 73–119.

Nebeker, F. (1995). *Calculating the Weather: Meteorology in the 20th Century*. Academic Press, London.

OECD (2009). Innovation in country risk management. *OECD, SwissRe & Oliver Wyman*, Paris [Online]. Available at: https://www.oecd.org/futures/Innovation%20in%20Country%20Risk%20Management%202009.pdf [Accessed 13 June 2020].

Palmer, T.N. (2002). The economic value of ensemble forecasts as a tool for risk assessment: From days to decades. *Quarterly Journal of Royal Meteorological Society*, 128 (581), 747–774.

Pigeon, P. and Rebotier, J. (2016). *Disaster Prevention Policies. A Challenging and Critical Outlook*. ISTE Press, London, and Elsevier, Oxford.

Porter, T.M. (1995). *Trust in Numbers: The Pursuit of Objectivity in Science and Public Life*. Princeton University Press, Princeton, NJ.

Pottier, N., Penning-Rowsell, E., Tunstall, S., Hubert, G. (2005). Land use and flood protection: Contrasting approaches and outcomes in France and in England and Wales. *Applied Geography*, 25(1), 1–17.

Power, M. (2004). *The Risk Management of Everything: Rethinking the Politics of Uncertainty*. Demos, London.

Richardson, L.F. (1922). *Weather Prediction by Numerical Process*. Cambridge University Press, Cambridge.

Rosa, H. (2013). *Social Acceleration: A New Theory of Modernity*. Columbia University Press, New York.

Rothstein, H., Huber, M., Gaskell, G. (2006). A theory of risk colonization: The spiralling regulatory logics of societal and institutional risk. *Economy and Society*, 35(1), 91–112.

Rothstein, H., Borraz, O., Huber, M. (2013). Risk and the limits of governance: Exploring varied patterns of risk-based governance across Europe. *Regulation and Governance*, 7(2), 215–235.

Thompson, P.D. (1984). A review of predictability problem. In *Predictability of Fluid Motions*, Holloway, G. West, B.J. (eds). American Physics, New York.

Tracton, M.S. and Kalnay, E. (1993). Operational ensemble prediction at the national meteorological center: Practical aspects. *Weather and Forecasting*, 8(3), 379–398.

UNISDR (2006). *Global Survey of Early Warning Systems*. UNISDR, Geneva.

Vinet, F. (2017). *Floods 2: Risk Management*. ISTE Press, London and Elsevier, Oxford.

Weber, M. (1992). *The Protestant Ethic and the Spirit of Capitalism*. Routledge, London.

8

Toward a New Security Deal? Reflexive Modernity, a Complex Turn and Shift to Uncertainty

Magali REGHEZZA-ZITT
École normale supérieure – PSL, LGP-Meudon (UMR 8591), Paris, France

8.1. Introduction

Strictly speaking, the Anthropocene refers to a new geological era (Crutzen and Stoermer 2000) defined by the fact that humanity has become a geological agent able to disrupt the planetary system through its actions and interventions on the biophysical environment. In addition to the controversies mentioned in the introduction to this book concerning the temporal demarcation of this geological era, the term can be interpreted as a new narrative (some call it a "metaphor", Garcia 2015) of the relationship between human societies and their environment. The Anthropocene then becomes a moment in Western thought, where the dualism between nature and culture is profoundly challenged (Kalaora 2007) and modernity turned reflexive (Beck *et al.* 1994) (see Chapter 6). The Anthropocene is also described as a historical event, concomitant with the globalization of economies, the complexification of sociotechnical systems and the planetary turning point (Chakrabarty 2015; Reghezza 2015).

The Anthropocene is closely linked to questions of environmental insecurity, which refer to the human condition and the way we dwell on the Earth. It is characterized by the growing perception of a significant increase in human vulnerability (Dalby 2015). The major disasters that have marked the beginning of the second millennium are characteristic and revealing of the "Anthropocene event"

coordinated by Julien REBOTIER. © ISTE Ltd 2021.

(Bonneuil and Fressoz 2013). On the one hand, so-called "natural" disasters involve high-intensity physical hazards and remind human societies of the existence of the planet through its destructive manifestations. However, these disasters cannot be explained solely by the action of "nature", whatever the meaning of this term. The damage is largely the outcome of the exposure and vulnerability of societies, which are rooted in current and inherited demographic, economic or political dynamics. So-called "natural" disasters thus perfectly illustrate the idea that disasters are manmade (O'Keefe *et al.* 1976; Chambers 1989). On the other hand, many disasters are now the result of hybrid hazards, produced by interactions between initially purely physical processes and human artifacts. The Fukushima nuclear accident in 2011, which is an example of a NaTech hazard (a combination of so-called natural and technological risks), is exemplary in this respect.

Through the idea of "threat", the Anthropocene questions the security project of modernity which, in tension with the imperative of freedom, ensures stability and protection for citizens, under the authority of the government. Whether they involve the planetary environment and the "common home" that is the Earth and/or its inhabitants, human and non-human, these threats have the particularity of being global, that is to say, both on a worldwide scale and systemic.

The use of the term "threat" is not neutral from an epistemological, social and political point of view. It accompanies the rediscovery of an uncontrollable uncertainty (Reghezza *et al.* 2019). In a modernity that has made risk an instrument of uncertainty control and governmentality (Foucault 2004b), the awareness of the existence of invisible, unpredictable and irreversible threats (Bourg *et al.* 2013), both transnational and trans-scalar, challenges the foundations of human security.

The following text, therefore, proposes to analyze the shift from a society of risk to one of threat, which seems to characterize the Anthropocene. This shift is the result of the globalization of hazards and vulnerabilities, which are themselves produced by the increasing complexity of socioecological systems.

8.2. From the globalization of threats to the resurgence of uncertainty

Financial risks, mass terrorism and climate change are the three archetypal figures of global threats. In this category, we can also classify the erosion of biodiversity, pandemics, dissemination of micro- and nano-plastics, acidification of the oceans, etc. Most of these threats are related to environmental security, that refers both to the security of the environment and the risks that the environment poses to human societies, due to the way we live on Earth. As such, these threats are linked to the Anthropocene. The global threats are often presented as new and characteristic of the 21st century, although the hazards that give rise to them are, for the most part, far from new.

8.2.1. *Globalization or* mondialisation *of environmental threats?*[1]

In the Anthropocene, many hazards are becoming global in the sense that they are upscaling and involve vast areas, or even the entire surface of the globe. Admittedly, its increasingly worldwide scale of initially localized disturbances is not something that has only occurred recently: the eruption of Tambora in 1815, that of Krakatoa in 1883 and the fissural eruption of Laki in Iceland in 1783 all had consequences on a global scale (Garnier 2009). The societies of the time, however, were unable to perceive their worldwide dimension, suffering only local, punctual and often time-lagged effects. The current media coverage of disasters and the progress of scientific knowledge now make it possible to grasp their worldwide extension almost instantaneously.

The globalization of threats cannot, however, be reduced to an evolution of individual and collective perceptions, which would result in the worldwide diffusion of information. We can indeed observe a scalar leap in the effective extent of threats, which now extend to the surface of the globe. Exposure is becoming ubiquitous, not so much because of a change in the scale of hazards as because of a change in the scale of their potential impacts, and through them, of vulnerability.

In the case of the above-mentioned eruptions, the transition to a worldwide scale results from the very nature of the dangers. These are diffuse hazards, which affect very large areas, spreading from near to far. This type of hazard and the dynamics of diffusion that it implies still exist: pollution clouds, oil slicks, certain epidemics.

But we can also see another dynamic of propagation, which leads to a rapid diffusion of hazards, over very large areas, with domino effects that lead to systemic crises, with varying periods of disruption and upheaval of landmarks and routines. Michel Lussault, referring to the eruption of the Eyjafjöll volcano in Iceland in 2010, spoke of a "globalizing crisis" to describe events that arise "from a spatially isolated fact that suddenly and contextually presents a problem, and whose problematic elements are diffused in the multi-fold networks of global hyperspace" (Lussault 2013, p. 223). If we take the example of this eruption that paralyzed part of the world's air traffic in 2010, we are dealing with a very localized hazard with a

1 Here we use both the term globalization and the French term *mondialisation* to denote different concepts. While globalization refers to the growing and complex interdependence of populations, markets and economies, due in part to increased deregulation and the advent of capitalism as the dominant global ideology, *mondialisation* denotes the idea of a transition to a worldwide or global scale.

physical phenomenon that was not at all extraordinary in terms of intensity, as the gas and ash projections never reached levels comparable to those of the major eruptions previously mentioned. The crisis was initially produced by the interaction of a physical process with sociotechnical systems. It was, therefore, part of the nature–culture hybridization that, in its trans-scalar dimension, characterizes the Anthropocene. The environmental hazard also provoked a systemic crisis, on a worldwide scale, with multiple impacts (economic, financial, social, political, etc.).

The global health crisis related to the spread of SARS-CoV-2 (COVID-19) in 2019 is also an example. The birth of the virus is linked to the interaction of humans and non-humans, with animals serving as reservoirs and intermediate hosts. This is a zoonosis, an infection whose agents are transmitted from animals to humans. Once again, we see the complex interaction between nature and human societies. The epidemic, initially localized in the Chinese province of Hubei, spread thanks to the vectors of international mobility (air transport in particular) to the entire planet. The progression of the epidemic clearly reflects the degree of integration into globalization and the *mondialisation* of international mobility, showing, at least initially, the margins and those excluded from the process. The health crisis is causing major economic, financial, social, political and geopolitical disruptions at all levels (regional, national, continental and global).

In the two examples cited, the diffusion on a worldwide scale results from a contagion due to the effect of contiguity, of functional proximity, and no longer of physical proximity. The scaling up of the disturbances is thus not, or not only, related to the nature or to the transformation of the sources of danger. Both crises affect the whole world and are fed by the vulnerabilities created by globalization, that is, the increase in interactions between individuals, societies and territories and the resulting complex interdependencies. These crises are becoming systemic because the disturbances affect all the components of the globalized world system beyond the nature of the initial hazard itself.

The systemic crises that characterize the Anthropocene thus reveal the way in which globalization transforms the threats and disasters that result from it. By intensifying interactions between individuals, activities and territories, and by giving them a worldwide and transnational character, globalization is at the origin of a contraction of space–time, which explains the rapid spread of threats to the point of causing systemic disturbances on a planetary scale. We must, therefore, distinguish between the *mondialisation* of threats (surface extension) and their globalization (systemic, transnational and trans-scalar, nonlinear character). In the globalized world of the Anthropocene, disturbances threaten systems of systems and are propagated because of multiple interdependent relationships, on various scales of time and space. The globalization of threats is thus only the expression of new forms of vulnerability, which arise from the complexification of social and economic organizations, which moreover interact permanently with anthropized biophysical systems, themselves complex. Complexity may be related to the number of components of the system under consideration and the intensity of their interactions, or to the interlocking of subsystems within the encompassing system, but in all cases, it produces uncertainty (Pigeon 2012). Thus, the so-called *mondialisation* of threats – the extension of exposure to the entire ecumene – is generally only the geographical translation of their globalization, that is, of the possible systemic contagion through domino effects and feedbacks, of the consequences of the initial disturbance (Reghezza *et al.* 2015). In other words, the globalization of threats is not just a scaling up in hazards; it is a change in the nature of crises.

8.2.2. *Global threats, spatial and temporal telescoping*

The first characteristic of global threats is that they are no longer of a multiscalar dimension but rather a trans-scalar one. What happens at the micro-level (the so-called local level) partly arises from what happens at the macro-level (the so-called global level), while having retroactive effects on the latter (Djament-Tran 2015). We thus observe a telescoping between spatial scales. Global hazards are "simultaneously local and global" (Smouts 2003). Very localized hazards can have global impacts, while global threats can have very localized consequences. In Swyngedouw's words, "the 'local' and the 'global' are deeply intertwined... In short, the local and the global are mutually constituted, or so it seems" (Swyngedouw 1997, p. 137).

The globalization of threats also raises the question of their temporality. It requires taking into account multiple time steps, which is generally reduced to the question of a temporal dilation of processes and their consequences, and to the problem of the articulation of short, medium and long periods of time. Just as the spatiality of the global cannot be reduced to a question of surface expansion, global time cannot be limited to a simple time dilation. Climate change, for example, requires us to look at time steps of several decades, or even several centuries, in order to understand the multiple outcomes of global warming. But beyond the need to re-inscribe the understanding of mechanisms and processes in the long term, the systemic nature of disturbances requires taking into account reactions, interactions, retroactions, at the scale of the components of the Earth system (ocean, cryosphere, biosphere, human societies, etc.), whose temporalities are not necessarily synchronous and with threshold effects that are difficult to measure.

Global threats thus mobilize multiple and nonlinear temporalities that reflect the rhythmic shifts in the deployment of cindynogenic dynamics: the multiple components of socioecological systems evolve at various speeds, with accelerations or, on the contrary, slowdowns, and longer or shorter latency times, so that multiple temporal shifts and "desynchronization" effects are observed (Djament-Tran 2015). These desynchronization effects increase the overall complexity and uncertainties, so that unpredictability becomes the rule (Reghezza *et al.* 2019).

8.3. From risk to threat: when complexity produces uncertainty

The globalization of risks arises from complexity. The latter produces uncertainties that do not (or no longer) fall within the scope of "risk-taking", in other words, that escape the category of risk, as it was constructed by the *Modernes*[2] precisely to circumscribe uncertainty. The shift from "risk" to "threat" can thus be interpreted as the resurgence of an uncontrolled uncertainty, which arises from the general complexity of the globalized world of the Anthropocene.

This does not mean that risk taking has become totally inoperative, that it has disappeared from our societies or that it no longer informs our way of seeing things. It does imply, however, that the way we perceive, define or use "risk" is changing.

8.3.1. *Uncertainty, risk-taking and modernity*

The expression "risk society" (Beck 1986) describes the fact that risk has become a foundation of the societies in which we live. We say that it has become that because it was not always the case. To speak of a "risk society" implies first of all defining risk not as a "finite category of phenomena" (this is what geography proposes when it describes risk as a situation of danger caused by the co-presence of a hazard and vulnerable stakeholders), but as "a mode of representation and a posture" (Peretti-Watel 2005) characteristic of modernity. Anthony Giddens, for example, asserted that risk (or, more precisely, the "culture of risk") was "a fundamental cultural aspect of modernity, through which awareness of the risks involved becomes a means of colonizing the future" (Giddens 1994, p. 244).

Risk has become for the *Modernes* the way to represent the world as uncertain, because "risk constitutes our *episteme*. It is at the heart of the relationship of individuals to themselves (morality), of individuals to nature (epistemology), of the relationship of individuals to each other (political anthropology)" (Ewald and Kessler 2000, p. 68). Risk is, therefore, both a cognitive category and a tool for dealing with uncertainty (see Chapter 7).

2 Here, we use the French concept of the *Modernes* to refer to populations rooted in modern tradition and philosophy, as opposed to post-modernity. It is worth noting that this is a Western-centered concept.

It is not a matter of saying here that uncertainty is an invention of modernity or that premodern societies had not developed answers *to deal with uncertainty*. For example, they resorted to superstition, to beliefs, to religions, to protect themselves from uncertainty, which they associated with metaphysical forces (God, the devil, etc.).

In modern societies, uncertainty is dealt with on the basis of "the three institutional pillars of Modernity – which are Science, the State and the Market" (Bourg *et al.* 2012). The State, through its representatives, will ensure the protection of citizens; Science will contribute to the eradication and, failing that, the control of threats; the Market will offer a model for managing risks based on insurance (Boudia 2013). For the *Modernes*, "risk-taking" is thus a rational response to the existence of randomness, unpredictability and the unknown. In contrast, uncertainty refers to what remains unknown, unknowable or unpredictable (Callon *et al.* 2001). In fact, in everyday language, risk and uncertainty can be synonymous, since risk includes an element of indeterminacy which is part of the uncertain. But in all rigor, "risk" and "uncertainty" cannot be interchanged.

The distinction between risk and uncertainty was developed at length at the beginning of the 20th century in the field of economics (Knight 1921; Keynes 1937). It gradually left this field and was taken up by many authors who defined uncertainty as the part of the indeterminacy of an event that remains irreducible because it cannot be calculated (Kermisch 2012). Risk then refers to the possibility of subjecting the hazard to the calculation of probabilities, knowing that the latter are not always "objective" or "frequentist" probabilities, that is, probabilities calculated *a posteriori* from series of systematic observations that make it possible to define a frequency of occurrence for a given event. Conversely, uncertainty, which can be more or less radical (i.e. measurable), refers to a situation where it is impossible to subject the hazard to the "standard space of calculation" (Bourg *et al.* 2013). We thus speak of risk when a hazard is "well identified, associated with the occurrence of an event or series of events, perfectly describable" (Callon *et al.* 2001) and of uncertainty when the event or its consequences are indeterminate by nature and not because of the unpredictability of the triggering event.

This opposition between uncertainty and risk can finally be assimilated to an operation of translation that marks the limits of the risk society: while "the term 'risk' is an attempt to make the unpredictable effects of our societal decisions predictable and controllable" (Beck 2003, p. 29), uncertainty refers to what cannot "be put at risk". To speak of uncertainty expresses the inability to convert the unknown or indeterminate into objective knowledge that can then allow for a "rational" choice (implications of such an inability for action are explored in Chapter 9). In this sense, uncertainty is what escapes the risk society because it escapes the cognitive and discursive tools that the latter has forged and which are the foundation of its own existence and security.

8.3.2. *Risk and complexity*

Numerous works in sociology (Beck 1992; Ewald 1986; Giddens 1994) from the turn of the 1990s converged on the idea that "risk appears as the counterpart of an increase in complexity" (Nicolet 2010, p. 11). Risk management is interpreted as the "materialization of an 'all-out complexity management' that does not dare to say its name" (Pesqueux, 2010, p. 7).

As Olivier Borraz notes:

> The notion of risk [...] offers a convenient way to deal with the complexity that surrounds us. In this sense, it is akin to an unlikely compass with which to venture into a space populated by uncertainties. (Borraz 2008, p. 11)

Complexity produces uncertainty. This is illustrated by the case of threats affecting certain "complex" systems, including the global risks of the Anthropocene. The greater the complexity, the more the feedback effects multiply and lead to forms of attenuation or, on the contrary, amplification of disturbances, which will explain the shifting of effects in time and space, and increase unpredictability (Reghezza *et al.* 2015).

However, the uncertainty produced by complexity has a singularity. Uncertainty can be similar to a conjunctural ignorance linked to a given and relative state of knowledge and/or of the means of producing this knowledge: for example, it can result from the absence of sufficient statistical series to establish a return frequency for a given hazard. It can also be produced by a widening of the field of knowledge that brings to light new problems of which we were previously unaware, new unknowns which it is not possible to resolve immediately in relation to the state of knowledge and scientific and technical tools. The uncertainty inherent in complexity is different, because it does not result from a state of affairs that could be changed: the non-predictability is structural (Dupuy 2002, pp. 131–132).

We must then ask ourselves whether the category of risk forged by the *Modernes* is capable of taking into account this "radical" uncertainty and whether risk can continue to play the role of "compass" described by Olivier Borraz. To put it another way, if risk is a way of mastering uncertainty by giving it a measurement, the irruption of the latter in contemporary scientific and political discourse could indicate the structural incapacity of contemporary societies to translate into risk the events that threaten our societies. This assertion calls for some precautions. As Chalas *et al.* (2009) rightly noted, many situations still fall under the heading of classical risk-taking. In other words, a distinction must be made between "radical" uncertainty, which escapes being "put at risk" because it cannot be parameterized or calculated, and uncertainty that is the result of cyclical ignorance and which will eventually be put at risk.

8.3.3. *Global shift, uncertainty and reflexive modernity*

If we admit that risk is constitutive of our relationship to the world, it is not just any world: the World of the Anthropocene, whose biophysical dimension is radically transformed by human activity.

The "risk society" is indeed confronted with a second phase of modernization (Latour 2014) in which modernity becomes "reflexive" (Beck *et al.* 1994). The first modernity was based on the values accorded to progress, science and technology and rationality, without these values being questioned. The second modernity is, on the contrary, marked by the questioning of its foundations, because technological progress and scientific innovation are generators of uncertainty, which are at the origin of such threats that the very existence of humankind is henceforth in peril (one thinks here of the risks linked to the proliferation of nuclear weapons, for example, which fueled the imagery of the Cold War without breaking with the primacy of the techno-scientific paradigm). In so doing, "Modernity becomes a problem in itself because it reflects itself as a risk" (Kessler 2002), hence the idea of reflexivity. In this framework, in the first modernity, the risk meant the control of uncertainty and, in the second modernity, uncertainties escape the control of the sciences and the techniques which gave rise to them. The "risk society", therefore, changes in meaning, because risk no longer has the same meaning.

In the Anthropocene, this reflexivity is moreover inseparable from the awareness of planetary finitude. As Lemarchand points out:

> Uncertainty has paradoxically intruded into societies that have chosen technical progress precisely in the hope of mastering nature and reducing randomness and indeterminacy. (Lemarchand 2013, p. 127)

It is in this sense, moreover, that the Anthropocene is inseparable not only from the reflection on modernity but also on its corollary, financial and economic globalization. The Anthropocene seems to see the failure of the modern project, which aims at the emancipation of individuals and the overcoming of the precariousness of human existence. It is now widely accepted that the dominant model of economic development, which bases individual and collective progress on the market economy and growth, has not succeeded in eradicating poverty, while causing irreversible damage to the environment, whether local or global. Dalby has shown, for example, how the global financial crisis of 2008 revealed the vulnerability of people around the world to economic hazards and reminded us that the state of extreme poverty of billions of people was aggravated by the removal of some of the social security mechanisms of the welfare state. At the same time, he points out that the improvement in the living conditions of current generations compared to previous generations is real, but that it has been achieved through the exploitation and artificialization of the planet, so that modern life is directed and shaped by the global economy (Dalby 2015). If we add that the consequences of environmental threats first affect the poorest individuals and communities, we understand that the Anthropocene leads to questioning the modern project at its very foundations.

8.4. The Anthropocene or the twilight of the modern project?

8.4.1. *Risk and modern governmentality*

In the modern State, risk is no longer associated with an event, but with a situation in which an action takes place (the action of risking). In this way, risk becomes not only the object of politics, but a guide of governmental rationality.

Risk remains an object of government. Those in power will have to define which risks must be fought – potential events that threaten the pursuit of individual interest – and which risks must be encouraged – situations of uncertainty that allow the exercise of freedom to act and choose.

Ulrich Beck pointed out the role of risk in the social contract between modern rulers and the ruled:

> The passage from the paralyzing fear of dangers to the calculation of risks is an essential component of the epic of modern times. In Europe, this triumphal march culminates in the development and organization of the security state and the welfare state, which bases its legitimacy on its capacity to protect its citizens against dangers of all kinds. This implies making all dangers calculable and controllable, and thereby making the danger disappear by transforming it into a risk. (Beck 2001, p. 6)

The "security state" is thus the *raison d'être* of the rule of law and legitimizes the exercise of power over the governed. In so doing, Beck shows the extent to which the logic of insurance has become part of the construction of the modern state. The welfare state functions as a collective insurance against social risks such as illness, old age, industrial accidents and unemployment, a function that nevertheless calls for a shift from individual foresight to collective protection. As long as the threat is thought of as a fatality, an act of God or the devil (in premodern societies), uncertainty belongs to the individual sphere. From the moment it is secularized, the threat becomes a risk and uncertainty becomes controllable, because it can be measured. It is then possible to move on to the idea of individual (insurance) and above all collective protection. This collective protection is both the attribute and the duty of the State, which becomes a "provider of protection" in the face of social insecurity (Castel 2003).

But risk is more fundamentally an instrument of governmentality and the specification of risk as a calculable risk plays a major role. Cost–benefit analysis, which presupposes the ability to quantify the threats, makes it possible to arbitrate between the different measures to be introduced and becomes a decision-making tool. However, it is necessary to dissociate calculable risk as a tool at the service of economic rationality, which serves to define and legitimize choices to act in a situation of uncertainty, and the regulation, management and control mechanisms of individual and collective social behavior, which are based on this tool without being reduced to it. Thus, those in power will define populations judged to be at risk: on the one hand, populations potentially dangerous to themselves and to others who it will try to control (prison, psychiatric hospital, judicial or social control) to protect society, and on the other hand, vulnerable populations, that is, in a situation of danger, who it will try to protect (children, elderly people, single women). In both cases, calculable risk is a tool at the service of a governmental technique: the authorities will rely, for example, on the probability of recidivism to treat delinquency or on the probability of negative side effects to authorize or administer a particular treatment. The calculable risk then allows the control of bodies. This is the birth of "biopolitics" (Foucault 2004a).

By relying on calculable risk, the modern government seeks not so much to limit or constrain as to bring out, develop, stimulate, orchestrate, and accompany the forces that it subjects, that is to say, it will, *in fine,* encourage risk-taking, this time in the sense of actions whose outcome is uncertain. By guaranteeing the security of the governed, the government allows the exercise of freedom, understood as that which allows each person to seize the opportunities of uncertainty.

In the end, for the *Modernes*, risk "defines a principle of politicization, a principle of the institution of politics, a little like the contract could be in the philosophy of natural law" (Ewald and Kessler 2000, p. 56). Risk becomes an instrument of governmentality at the cost of a shift in meaning that makes it pass from a threat that must be prevented, or failing that, administered (the logic of the contract), to a positive opportunity that informs society in its functioning and its aspirations.

8.4.2. *Risk and political economy*

Risk as a principle of politicization illustrates more broadly the shift to political economics, that is, the fact that governmental rationality is modeled on that of the market economy. The government transfers economic models and references into public action, so that economic rationality becomes a principle of government. Soraya Boudia explains in particular that although the technologies that aim to manage risks are an assembly of extremely diverse practices and rationalities, they call upon a common base of knowledge that combines economic calculation, optimization and behavioral sciences, based on the notion of rational choice, which postulates that individuals always act according to a utilitarian logic and that an action is considered rational from the moment its benefit is greater than its cost (Boudia 2013).

In the modern state, the reconciliation of liberty and security remains at the center of the problem of government, but the balance between the two has shifted, which changes the meaning of the social contract. Indeed, through political economy it is the principle of absolute freedom to pursue one's own interests that is asserted, which implies taking risks. Government no longer resides in constraining the freedom of the governed, but in manipulation of the interests of each individual, the social contract guaranteeing that the individual will be exposed as little as possible to the dangers that threaten the pursuit of his interest (Foucault 2004a). If we take up Foucault's analysis, whereas at the beginning of modernity, reason aspired to govern men through the intermediary of regulations and norms, the State's intervention in society will henceforth boil down to producing conditions that will allow the market to play its role as regulator. Translated into terms of risk, this means managing threats to the security of individuals, society and transactions, threats in the sense that they would prevent everyone from taking the risks they deem appropriate.

By becoming the principle of governmental rationality, political economy results in the government "engaging in a mechanism in which it will have to arbitrate the freedom and security of individuals at every moment around the notion of danger" (Foucault 2004a, p. 15). From then on, governing consists first of all of governing *by* risk, insofar as those in power will interpret and think of reality as a risk, and where it will arbitrate between the risks that will have to be managed (events, but also certain categories of the population) and those that will have to be optimized (situations of uncertainty that allow the pursuit of interest). Uncertainty is then valued on condition that it can be controlled, which implies both controlling certain events in order to create conditions favorable to risk-taking *and* encouraging the freedom to take risks by imposing the idea that it is possible to choose rationally in a situation of uncertainty, which in both cases leads to assimilating risk to calculable risk and devaluing radical uncertainty.

8.4.3. *Anthropocene uncertainty: failure or paradoxical reinforcement of modern governmentality?*

In the Anthropocene, the resurgence of uncertainty and the shift toward a "threat society" undermine the foundations of modern governmentality. Indeed, radical uncertainty makes it impossible, by definition, to put oneself at risk. If we consider that the modern State derives its legitimacy from its capacity to guarantee security by controlling uncertainties, the irruption of an uncertainty that cannot be parameterized – and by the same token, cannot be controlled – does not so much signify the failure of technology to prevent disasters, nor even the inability of science to foresee them, but rather the incapacity of the power in place to dominate the future sufficiently to ensure the security of citizens and the community, and thus to fulfill the mission that legitimizes its existence. Radical uncertainty should be unthinkable in the modern State: to show it would be to admit its failure and to recognize the inanity of the security guarantees on which it is based and which legitimize it.

However, we are witnessing a paradox: in the Anthropocene, radical uncertainty leads to the consolidation of the government of societies through risk. For the *Modernes*, the parameterization of uncertainty had constituted an instrument of governmentality inscribed in a wider security system. Uncertainty was accepted as a horizon of ignorance to be overcome, a temporary unknown produced by the momentary lack of knowledge. The radical uncertainty inherent in the complexity of the globalized world of the Anthropocene is irreducible to this representation of the uncertain. It escapes, by its very nature, the modern device of risk-taking. Yet, far from destroying this device, radical uncertainty, perceptible through global environmental threats, reinforces government through risk and strengthens the biopolitical dimensions of governmentality, notably through injunctions to adapt and resilience (Rufat 2011).

8.5. Adaptation and resilience: transformative promise or conservative revolution?

Resilience and adaptation are initially descriptive scientific concepts used in multiple disciplinary fields (Reghezza-Zitt and Rufat 2016). Their operational success has been widely studied. Many works show that resilience has been explicitly linked to the uncertainty produced by the complexity of socioecological systems. Berkes insists, for example, that it is impossible to predict the behavior of a complex system undergoing a disturbance, arguing that the response of the latter will be nonlinear and that the consequences of the initial disturbance will be beyond the proportions of the latter. Insofar as, he writes, social systems, like ecological systems, are complex systems, we can deduce that uncertainty becomes a fundamental fact of contemporary societies. For Berkes, complex systems undergo changes that are both linked to their co-evolution and to the transformations that affect the planetary system. These changes are unpredictable. Global warming is the archetype. It is therefore necessary, in this context, to develop approaches that make it possible to "cope with change that cannot be predicted" (Berkes 2007).

This is where adaptation and resilience come in, not as scientific concepts, but as political discourses and normative injunctions. Resilience, defined here as the ability to overcome a disturbance, whether a slow stress or a sudden shock – in this case the global changes that characterize the Anthropocene – arises from the adaptive capacities of the systems (individual, technical system, territory, etc.) considered (Reghezza-Zitt and Rufat 2016). Resilience thus becomes a goal to be achieved through the development of the adaptive capacities, that is, of transformation, of societies and individuals. It is no longer a question of mitigating hazards, nor even of reinforcing resistance to shock (a reinforcement that constituted the basis of vulnerability reduction policies), but of developing a set of properties, faculties, qualities, both individual and collective, that will enable the issue to transform itself in order to absorb the shock resulting from the changes that affect it internally and externally.

Resilience thus allows for a shift from a frame of reference of risk that relied on hazard mitigation and vulnerability reduction to the frame of reference of uncertainty that assumes adaptation. Political power gives up on dealing with uncertainty as such, through action on the causes that produce uncertainty and the structural factors of vulnerability (Wisner 2016). Populations accept this renunciation, which is no longer perceived as a failure of government, but, on the contrary, as something "normal". Chandler thus describes the tour de force that leads to acceptance of the idea that the most urgent task of states is now "empowering the subject to make better life choices" (Chandler 2013, p. 219), instead of protecting the vulnerable.

Within this use of resilience, the principle of political economy – risk is the ontological condition of man – is pushed to its paroxysm. Both the political and moral responsibility for dealing with uncertainty is transferred from government to communities and individuals, in the name of a capacity for self-organization that can be interpreted as "the inculcation of a morality of effort and independence, and of an attitude appropriate to the functioning of the market" (Felli 2014, p. 107).

The injunction to adapt thus bears the seeds of a societal project that aspires to nothing less than the transformation of individuals, who, from victims assisted by state intervention, become, in a way, masters of their own destiny. It is no longer the state's control over external threats, but the "self-government" (Chandler 2013, p. 212) of vulnerable populations and individuals that becomes the principle of government. The object of government is no longer so much the security of populations, but the populations themselves (Dalby 2013). For Felli, for example, "the promotion of adaptation is now essentially the production of a new ethic focused on transforming the perceptions, attitudes and behaviors of populations considered vulnerable to environmental change" (Felli 2014, p. 109).

The same reasoning can be applied to resilience, which is the counterpart of adaptation. For Welsh, resilience is a governing technique whose goal is "to enable individuals, institutions, eco-systems and economies to be responsible for transforming themselves in the face of a world of contingency whilst also increasing their resistance to exogenous and internal shocks by limiting the potential of events to provoke change" (Welsh 2014, p. 19). For Pugh, resilience is even "a biopolitical response to the contemporary idea of complex life" (Pugh 2014, p. 316).

Adaptation and resilience are thus strongly critized because they promote the integration of the neoliberal norm that the resilient subject is adaptable to any form of uncertainty, and even capable of transforming uncertainty into an opportunity to improve his or her condition. The critique of resilience and adaptation is not about their necessity in a context of global threats whose consequences are already perceptible, but about the use that can be made of them to legitimize inaction, especially in terms of mitigation, and to transfer the costs of this inaction onto the most fragile, the excluded, the dominated, while at the same time making the root causes of their vulnerability invisible.

8.6. Conclusion: an unprecedented collective challenge

The awareness of *planetary boundaries* and global environmental threats, such as global climate change or the erosion of biodiversity, has reminded humanity of its condition as a vulnerable species imprisoned in a biophysical environment that, while it provides the resources necessary for its survival and development, can also produce risks that could destroy it. As Boucheron reminds us, in the Anthropocene, "humans know that they are condemned to live on a limited surface because they realize that their actions can modify the envelope" (Boucheron *et al.* 2009). In this respect, disasters can be considered as particular times when the vulnerability of societies in the Anthropocene, including the most technically and economically developed, is dramatically actualized (Reghezza *et al.* 2019). Vulnerability reflects *in fine* the awareness of finitude, both individual and collective. Global environmental threats and "natural" disasters fuel the awareness of a community of destiny between the habitat and the inhabitant, a community of asymmetric destiny due to the contingency of humanity on the planet.

Faced with an ecological catastrophe involving the survival of the human race, a catastrophe that has been declared inescapable, there is a great temptation to slide from accepted and collectively negotiated risk to the acceptance of a socialization of risk that places the burden of their security on individuals, and no longer on States (Reghezza *et al.* 2015). As Catherine and Raphaël Larrère observe, "if there is a new myth to study, it is that of the certainty of the coming catastrophe. What has changed [...] is the regime of fear" (Larrère and Larrère 2013, p. 200).

The problem is that a catastrophe is never certain, since between the risk, the realization of the hazard and the cataclysm, there remains precisely the uncertainty that allows very different attitudes, if not postures. As Bauman says so clearly:

> Fear is arguably the most sinister of the demons nesting in the open societies of our time. But it is the insecurity of the present and uncertainty about the future that hatch and breed the most awesome and least bearable of our fears. That insecurity and that uncertainty, in their turn, are born of a sense of impotence: we seem to be no longer in control, whether singly, severally or collectively – and to make things still worse we lack the tools that would allow politics to be lifted to the level where power has already settled, so enabling us to recover and repossess control over the forces shaping our shared condition while setting the range of our possibilities and the limits to our freedom to choose: a control which has now slipped or has been torn out of our hands. The demon of fear won't be exorcized until we find (or more precisely construct) such tools. (Bauman 2007, p. 26)

The prevailing catastrophism can be interpreted as a reflection of the inability to control (or master) uncertainty and to put it at risk. Faced with the impossibility of turning back the clock, there is a great temptation to rush forward (Rufat 2011): the injunction to adapt is a good example of this, since in its technicist reading, it is interpreted as an encouragement to continue along the same path of continuous progress in the sciences, allowing us to compensate for their negative effects in the medium and long term.

It is for this reason that Dominique Bourg can write that:

> Ulrich Beck's vision – which envisages the risk society as one of a reflexive modernization that is more concerned with the secondary effects of techno-scientific developments – must be questioned, and with it the discourses that accompany it. One can indeed fear that the function of the confession – confession of a lack of control over the overflows of the technosciences – is more in the order of accepting criticism of the reflexivity, in sum, a step more toward the "disinhibition" facing this new Modernity. (Bourg *et al.* 2013, p. 5)

And to add that:

> Risk suggests that humanity could extricate itself from the impasse into which it is sinking by small economic calculations and/or by some technological *deus ex machina.* (Bourg and Whiteside 2010, p. 116)

The question that arises is whether we can escape from catastrophism, and if so, at what cost? What is the moral and political cost of optimism in environmental matters? What is the social and human cost of global catastrophism?

8.7. References

Bauman, Z. (2007). *Liquid Times: Living in an Age of Uncertainty*. Polity Press, Cambridge.

Beck, U. (1992). *Risk Society: Towards a New Modernity*. Sage Publishing, London and New York.

Beck, U. (2001). La dynamique politique de la société mondiale du risque. *Idées pour le Débat*, 1, 1–22.

Beck, U. (2003). La société du risque globalisé revue sous l'angle de la menace terroriste. *Cahiers internationaux de sociologie*, 114(1), 27–33.

Beck, U., Giddens, A., Lash, S. (1994). *Reflexive Modernization: Politics, Tradition and Aesthetics in the Modern Social Order*. Stanford University Press, Stanford, CA.

Berkes, F. (2007). Understanding uncertainty and reducing vulnerability. Lessons from resilience thinking. *Natural Hazards*, 41(2), 283–295.

Bonneuil, C. and Fressoz, J.-B. (2013). *L'Événement Anthropocène. La Terre, l'histoire et nous*. Le Seuil. Paris.

Borraz, O. (2008). *Les politiques du risque*. Presses de Sciences Po, Paris.

Boucheron, P., Loiseau, J., Monnet, P., Potin, Y. (eds) (2009). *Histoire du monde au XVe siècle*. Fayard, Paris.

Boudia, S. (2013). La génèse d'un gouvernement par le risque. In *Du risque à la menace*, Bourg, D., Joly, P.-B., Kaufmann, A. (eds). PUF, Paris.

Bourg, D. and Whiteside, K.H. (2010). *Vers une démocratie écologique : le citoyen, le savant et le politique*. Le Seuil, Paris.

Bourg, D., Joly, P.-B., Kaufmann, A., Lemarchand, F., Mormont, M., Pécaud, D. (2012). Comprendre et maîtriser les risques techniques et environnementaux : aller au-delà du risque ? *[VertigO] La revue électronique en sciences de l'environnement*, 12(1) [Online]. Available at: http://vertigo.revues.org/12152 [Accessed 27 May 2020].

Bourg, D., Joly, P.-B., Kaufmann, A. (eds) (2013). *Du risque à la menace. Penser la catastrophe*. PUF, Paris.

Callon, M., Lascoumes, P., Barthe, Y. (2001). *Agir dans un monde incertain : essai sur la démocratie technique*. Le Seuil, Paris.

Castel, R. (2003). *L'insécurité sociale : qu'est-ce qu'être protégé ?* Le Seuil, Paris.

Chakrabarty, D. (2015). The human condition in the Anthropocene. *The Tanner Lectures in Human Values*, 18–19.

Chalas, Y., Gilbert, C., Vinck, D. (eds) (2009). *Comment les acteurs s'arrangent avec l'incertitude*. Archives Contemporaines, Paris.

Chambers, R. (1989). Editorial introduction: Vulnerability, coping and policy. *IDS Bulletin*, 20(2), 1–7.

Chandler, D. (2013). Resilience and the autotelic subject: Toward a critique of the societalization of security. *International Political Sociology*, 7(2), 210–226.

Crutzen, P.J. and Stoermer, E.F. (2000). The "Anthropocene". *IGBP Newsletter*, 41, 17–18.

Dalby, S. (2013). Biopolitics and climate security in the Anthropocene. *Geoforum*, 49, 184–192.

Dalby, S. (2015). Anthropocene formations: Environmental security, geopolitics and disaster. *Theory, Culture & Society*, 34(2–3), 233–252.

Djament-Tran, G. (2015). Resilience: A question of scale. *Resilience Imperative: Uncertainty, Risks and Disasters*, Reghezza-Zitt, M. and Rufat, S. (eds). ISTE Press, London, and Elsevier, Oxford.

Dupuy, J.-P. (2002). *Pour un catastrophisme éclairé. Quand l'impossible est certain.* Le Seuil, Paris.

Ewald, F. (1986). *L'État providence*. Grasset, Paris.

Ewald, F. and Kessler, D. (2000). Les noces du risque et de la politique. *Le Débat*, 109(2), 55–72.

Felli, R. (2014). Adaptation et résilience : critique de la nouvelle éthique de la politique environnementale internationale. *Éthique publique*, 16, 101–119.

Foucault, M. (2004a). *Sécurité, territoire, population. Cours au Collège de France, 1977–1978*. Le Seuil, Gallimard, Paris.

Foucault, M. (2004b). *Naissance de la biopolitique. Cours au Collège de France, 1978–1979.* Le Seuil, Gallimard, Paris.

Garcia, P.-O. (2015). Sous l'adaptation, l'immunité : étude sur le discours de l'adaptation au changement climatique. PhD Thesis, Université Grenoble Alpes, Grenoble.

Garnier, E. (2009). Laki : une catastrophe européenne. *L'Histoire*, 343, 72–77.

Giddens, A. (1994). *Les Conséquences de la modernité*. L'Harmattan, Paris.

Kalaora, B. (2007). La mise sur orbite planétaire de la nature. La nature qui relie ou qui délie. In *L'émergence des cosmopolitiques*, Lolive, J., Soubeyran, O. (eds). La Découverte, Paris.

Kermisch, C. (2012). Vers une définition multidimensionnelle du risque. *[VertigO] La revue électronique en sciences de l'environnement*, 12(2) [Online]. Available at: https://journals.openedition.org/vertigo/12214 [Accessed 14 May 2020].

Kessler, D. (2002). Ulrich Beck et la société du risque. *Commentaire*, 100(4), 889–892.

Keynes, J.M. (1937). The general theory of employment. *Quaterly Journal of Economics*, 51(2), 209–223.

Knight, F.H. (1921). *Risk, Uncertainty and Profit*. Sentry Press, New York.

Larrère, C. and Larrère, R. (2013). Peut-on échapper au catastrophisme ? In *Du risque à la menace*, Bourg, D., Joly, P.-B., Kaufmann, A. (eds). PUF, Paris.

Latour, B. (2014). *Cogitamusb : six lettres sur les humanités scientifiques*. La Découverte, Paris.

Lemarchand, F. (2013). Pour dépasser le concept de risque. In *Du risque à la menace*, Bourg, D., Joly, P.-B., Kaufmann, A. (eds). PUF, Paris.

Lussault, M. (2013). *L'avènement du monde. Essai sur l'habitation humaine de la Terre.* Le Seuil, Paris.

Nicolet, J.-L. (2010). *Risques et complexité*. L'Harmattan, Paris.

O'Keefe, P., Westgate, K., Wisner, B. (1976). Taking the naturalness out of natural disasters. *Nature*, 260(5552), 566–567.

Peretti-Watel, P. (2005). La culture du risque, ses marqueurs sociaux et ses paradoxes. *Revue économique*, 56(2), 371–392.

Pesqueux, Y. (2010). Une perspective contemporaine du risque [Online]. Available at: https://hal.archives-ouvertes.fr/hal-00509685 [Accessed 27 May 2020].

Pigeon, P. (2012). *Paradoxes de l'urbanisation. Pourquoi les catastrophes n'empêchent-elles pas l'urbanisation ?* L'Harmattan, Paris.

Pugh, J. (2014). Resilience, complexity and post-liberalism. *Area*, 46(3), 313–319.

Reghezza, M. (2015). De l'avènement du Monde à celui de la planète : le basculement de la société du risque à la société de l'incertitude. HDR thesis, Université Paris 1-Panthéon Sorbonne, Paris [Online]. Available at: https://tel.archives-ouvertes.fr/tel-01255031 [Accessed 27 May 2020].

Reghezza-Zitt, M. and Rufat, S. (2015). L'adaptation en Île-de-France entre injonction et recyclage. Techniques et politiques de la société de l'incertitude. *Développement durable et territoires*, 6(3) [Online]. Available at: https://journals.openedition.org/developpement durable/11035 [Accessed 3 September 2021].

Reghezza-Zitt, M. and Rufat, S. (2016). *Resilience Imperative: Uncertainty, Risks and Disasters*. ISTE Press, London, Elsevier, Oxford.

Reghezza-Zitt, M. and Rufat, S. (2019). Disentangling the range of responses to threats, hazards and disasters. Vulnerability, resilience and adaptation in question. *Cybergeo: European Journal of Geography*, 916.

Rufat, S. (2011). Critique de la résilience pure [Online]. Available at: https://hal.archives-ouvertes.fr/hal-00693162 [Accessed 27 May 2020].

Smouts, M.-C. (2003). Un trou noir dans l'espace mondial : le risque environnemental global. In *Entre Kant et Kosovo*, Le Gloannec, A.-M., Smolar, A. (eds). Presses de Sciences Po, Paris.

Swyngedouw, E. (1997). Neither global nor local: "Glocalization" and the politics of scale. In *Spaces of Globalization: Reasserting the Power of the Local*, Cox, K. (ed.). The Guilford Press, New York.

Welsh, M. (2014). Resilience and responsibility: Governing uncertainty in a complex world. *The Geographical Journal*, 180(1), 15–26.

Wisner, B. (2016). Vulnerability as model, concept, metric and tool. *Oxford Research Encyclopedia* [Online]. Available at: https://oxfordre.com/naturalhazardscience/view/10.1093/acrefore/9780199389407.001.0001/acrefore-9780199389407-e-25 [Accessed 9 May 2020].

9

The Imperative of Practical Wisdom in the Face of Anthropocene Emergencies: The Case of Climate Change[1]

Florent CHAMPY

LISST, CNRS, Toulouse, France

9.1. Introduction

When we reflect on the attitudes to adopt toward Anthropocene threats, the concept of *phronesis*, which has a basis in the Aristotelian philosophical tradition[2] (Aubenque 1963; Broadie 1991) and translated to practical wisdom or prudentiality[3], can be illuminating. Using Knight's distinction between risks and uncertainties as a

1 This chapter benefited from discussions with Marie-Pierre Bes, Guillaume Carbou, Pascal Ducournau, Raphaël Kunstler, Jérôme Lamy, Alexandre Malley-Ernewein and Renaud Nougarol, as well as from reviews of an earlier version by Vincent Hecquet, Jérôme Lamy, Jean-François Orianne, Patrick Pigeon and Julien Rebotier. I would like to thank them all for their help.

2 This concept is also found in the work of the Stoics, Spinoza and Kant. But I will start here from Aubenque's commentary on the Aristotelian concept of *phronesis*. For a perspective on the historical variations in the use of this concept, see Delannoi (1995).

3 "Prudentiality" is a neologism constructed from the Latin root *prudentia*, which gave "*prudence*" in French and "*prudencia*" in Spanish. Practical wisdom is a better translation of the Greek *phronesis*, as "*prudence*" and "*prudencia*" have a connotation of pusillanimity that is absent from the concept of *phronesis*, as we shall see. But this Latin root is used in the construction of the English adjective "prudential".

coordinated by Julien REBOTIER. © ISTE Ltd 2021.

starting point helps us to understand its contribution (Knight 1921). Knight describes as "risks" those dangers for which we know how to measure a probability of occurrence (in particular random phenomena of a serial type, which can be forecast), while "uncertainty" characterizes situations for which this measurement is impossible, or even for which the reality of the danger is not proven (e.g. when the use of a new technology gives rise to a possible new danger).

In everyday language, however, "risk" refers to both risks in Knight's sense and non-probabilistic hazards. This confusion is also found in much of the relevant scientific literature. Thus, the "sociology of risk" was formed in particular following Beck's use of the expression "risk society" (Beck 1992), but it deals as much with uncertainties as with risks. However, even though it is sometimes difficult to make a concrete distinction[4] between the two, such a contrast would be all the more useful since the epistemic attitudes that enable facing probabilistic dangers and irreducible uncertainties are different.

When a danger is probabilistic, deliberation on the basis of the risks actually incurred is possible, at least in theory[5]. But this is impossible in situations of irreducible uncertainty. However, the latter may require action, because indications of a possible serious danger may appear before science is able to measure its probability, or even to confirm its reality. In addition, action is sometimes necessary before science can provide guidance on the most effective actions. An approach different from cost/benefit/risk assessments and the application of scientific knowledge is then required to guide action. But a lack of distinction between risks and uncertainties allows irreducible uncertainties to be understood as though they were quantifiable risks or, more generally, can cause actors to act in denial of the uncertainties they are confronted with. The first part of Beck's work on the risk society thus already described the efforts of many actors to mask uncertainty and ignorance, and to move forward as if they knew (Beck 1992). One difficulty is that the epistemic attitude adapted to situations of irreducible uncertainty is in fact a mystery for many actors and researchers.

The concept of practical wisdom precisely indicates an epistemic and moral attitude (it is, at the same time, a way of thinking and a virtue) making it possible to act in situations charged with irreducible uncertainties, by seeking to prevent both possible dangers and the potential damage caused by the solutions to these dangers

4 The difficulty comes from the fact that, even if they are probabilistic, risks bring about uncertainties.

5 There is a long way to go from scientific certainty to clear information for citizens and public debate, as illustrated by the success of the tobacco industry's efforts to instil doubt about the harmfulness of its products (Oreskes and Conway 2010).

themselves. Practical wisdom thus compensates for the limits of technical rationality, supported in particular by scientific knowledge or numerical evaluations[6]. It has its place in situations that Knight has qualified as uncertain. It is precisely because many uncertainties cannot be eliminated before acting that it is necessary.

Whether we are interested in climate change, pollution or the decline in biodiversity, environmental history abounds in examples of situations where practical wisdom (sufficient attention to the signs, at first tenuous and then increasingly clear, of possible danger) was lacking. But this relevance of the concept of practical wisdom for understanding past errors, to which I will return, says nothing about its relevance today. As uncertainties have disappeared concerning the reality, the seriousness and the anthropic origin of climate change, of the decline in biodiversity or of various forms of pollution, can the concept of practical wisdom still help us to analyze the attitudes to adopt when faced with these dangers?

I will take this question as a starting point to rethink the place that practical wisdom can still hold in the apprehension of environmental problems and in the answers to be brought to them. I will illustrate my point with key themes of industrial modernity: growth, technical progress, consumption and rationality. I will examine what is being said about it in public debates on the environment and I will also draw on two sets of works: (1) the writings of economists who, contrary to dominant discourse in their discipline, give a large place to the environmental question (Caminel *et al.* 2014; Banerjee and Duflo 2019; Laurent 2019); (2) research on the place of practical wisdom in modern Western societies, conducted on the basis of surveys of other objects – the precautionary principle (Champy and Lepiller 2016) and professions (Champy 2018a, 2018b; Kuehni 2019).

I will defend a fourfold thesis: (1) the paucity of practical wisdom is not an accident, but the main limitation of ways of thinking consubstantial with modernity. (2) criticism of the latter does not mechanically translate into an increase in practical wisdom; (3) prudential attitudes toward environmental dangers are as necessary now as in the past; (4) the errors that practical wisdom would allow us to avoid are no longer exactly the same as at the end of the 20th century, so the use of this concept to think about social conditions favorable to practical wisdom must be historicized.

I will proceed in four steps. The concept of practical wisdom is not common in the study of risks and disasters. It is even widely misunderstood in our societies. I will therefore begin by presenting it (section 9.2). I will then examine the reasons for

6 On the excessive reliance on numbers, which leads to quantifying what cannot be quantified, see Porter (1995). But practical wisdom is not opposed to the use of figures. Actually, it often relies on them. I will come back to this.

affirming that lack of practical wisdom is consubstantial with industrial modernity (section 9.3), and that the issue of practical wisdom is no longer exactly the same as it was a few decades ago (section 9.4). Finally, I will explain the type of errors from which the concept of practical wisdom can help to protect oneself, starting from three of its facets (section 9.5): the consequentialist dimension of practical wisdom, distrust of polarized beliefs and its relation to rationality and rationalism.

9.2. The main characteristics of practical wisdom

The concept of practical wisdom refers both to a way of thinking about action in situations charged with an irreducible uncertainty, and to a virtue, since this way of thinking aims at finding the action most likely to avoid damage to others, or at least the most severe damage[7]. Uncertainty is due to the complexity and singularity of the situation. In a situation with little singularity, habitual actions would be appropriate. Furthermore, if the situation was not complex, the mobilization of knowledge would make it possible to remove any uncertainty. A prudential approach implies adapting the action by rooting it in concrete issues and observations, and by rejecting what may seem obvious but has not been put to the test of the situation. It is opposed to habits and the mechanical application of rules or general knowledge, even if knowledge, rules and routines can help take into account certain aspects of the situation. Among the fields of action requiring a high density of practical wisdom, healthcare (Champy 2015), social work (Kuehni 2019), education, environmental policies and economic policies are worth mentioning.

The Aristotelian philosophical tradition provides a phenomenology of prudential attitudes that can guide the analysis of action in situations of uncertainty. Practical wisdom first concerns the view of the situation to which the action must respond. A global view must be sought in order to avoid the damage that could be caused by not taking into account one of its dimensions. Moreover, attention must be paid to signs, however slight, that a danger may be greater than it appears[8]. Practical wisdom is also deliberative, and deliberation often concerns not only the means to be used in the action, but also the ends to be pursued as a matter of priority. For example, futile medical care stems from a lack of practical wisdom: proceeding

7 Practical wisdom can also allow one to avoid damage to oneself. But, following Aubenque, this is not what interests me here.

8 In medicine, an array of symptoms suggesting one pathology sometimes refers to another, of which a discrete symptom may be a clue. The clinical skill consists of grasping which signs are actually significant.

mechanically from the norm of curative performance, it does not constitute the right action since it can bring only additional suffering. Thus, practical wisdom implies choosing one's own rule of action. A doctor must in general do everything to cure their patient, but if the absence of hope abolishes this rule, they must know how to change it. Prudential action is therefore political: the hierarchy of ends it sets for itself cannot be built on purely technical foundations (Champy 2018a).

Being opposed to both necessity and chance, practical wisdom is difficult to conceive of within the framework of the dominant conceptions of the world. It is opposed to necessity because it is valid when it is impossible to determine with certainty which action is best adapted to the situation. But it refuses to rely on chance: to seek the right action with wisdom is to recognize that not all actions are equal. Intuition and experience make it possible to overcome the opposition between necessity and chance. When science, technique and calculation leave room for uncertainty, prudential action supposes a remarkable capacity to grasp what should be done in spite of these uncertainties, to *feel* what the right action is. Practical wisdom is, therefore, not only a way of thinking and a virtue, but also a capacity for proper judgment. It cannot be taught or transmitted systematically; experience contributes to its development. Finally, in the absence of certainty, error is always possible and prudential action is therefore often accompanied by doubt. Being full of certainty and not being able to tolerate doubt is not conducive to practical wisdom. Practical wisdom is modest, which contributes to its lack of credibility[9] in societies where certainty, even at the cost of hasty simplifications, is valued, and where uncertainty is often denied. However, acting prudentially implies knowing how to conjecture and then, once decisions have been taken, assuming uncertainties and the risk of error, often with audacity and determination[10].

Thus, practical wisdom aims at acting as well as possible in a complex situation, taking into account the possible consequences of the various conceivable responses, even though the complexity of the situation makes these consequences uncertain. It is a form of consequentialism with strong requirements: to be wary of the consequences of one's own actions, without arguing for uncertainties to justify

9 This is why the term prudentia, which gives a connotation of pusillanimity, harms the idea it is supposed to serve, and "practical wisdom" is a better translation.

10 History, literature and film abound with examples of people whose prudence is coupled with extreme boldness. According to Aubenque, Aristotle described practical wisdom by observing the actions of Pericles. A biography highlights the uncertainties faced by this Greek citizen during his long political career (Kagan 1998). Joe Wright's film *Darkest Hour* (2018) shows how British Prime Minister Winston Churchill decided to repatriate the British army encircled at Dunkirk by the Germans on fishing and trading vessels.

inaction; to be wary of everything that could lead to mechanical, insufficiently thought-out actions (beliefs, habits, conformism, private interests, biases linked to specialization, etc.); to anchor the reflection in the actual situation, using all the means available to grasp its complexity (scientific knowledge, quantified evaluations, observations, etc.); not to fetishize some finalities of the action at the expense of other possible stakes. Let us now see how understanding these requirements helps us to think about industrial modernity and the damage that it has produced, which is part of the Anthropocene.

9.3. Industrial modernity and denial of the prudential imperative

A lack of practical wisdom is inherent in industrial modernity. To understand this, let us begin by looking at the origin of the threats that constitute the Anthropocene, in particular climate change. Warning signs appeared long before the reality and seriousness of these threats were clearly established, without leading to any concrete precautionary measures. After the emergence in the 1970s of the hypothesis of anthropogenic global warming, the Intergovernmental Panel on Climate Change (IPCC) was created in 1988, but little was done to redirect economies toward less carbon-based energy and more energy efficiency. There are many explanations for the lack of action: denial of climate change or its anthropogenic origin by some, focus on short-term issues (due to public action agendas and the way financial capitalism constructs economic value), or the tendency to deal only with problems for which solutions exist. Warnings about the planet's vulnerability have not prevented environmental degradation beyond the most pessimistic forecasts: States, in particular, have failed to pay attention to these warnings and to the dangers they signaled, and have therefore failed to act prudentially. How is this consubstantial with industrial modernity?

Let us take the list of prudential requirements presented at the end of the previous section, and confront it with criticisms of modernity, well-known and examined in several chapters of this book. The first requirement of prudential action is to consider all the consequences of one's actions. Yet capitalism has developed by externalizing the cost of the negative consequences of its activity: pollution, depletion of resources or even occupational diseases and accidents. While opening up rights to individuals, the construction of an insurance system exempts companies from responsibility for the consequences of their activity, for a low cost: this is the socialization of risks, without which industrial economies could not have developed as they have. Secondly, practical wisdom requires that we do not fetishize certain ends of action to the detriment of other dimensions. Both maximum profit-seeking by companies and the obsession with growth on the part of governments are forms of fetishization: these objectives surpass all others, to the point where they are not even considered. Practical wisdom also implies being wary of beliefs, the personal interests of actors or habits that may be obstacles to reflexivity about the consequences of action. However, both the search for profit and the search for growth are based on two beliefs that have contributed greatly to ecological destruction and predation. The first, as old as Mandeville's *The Fable of the Bees* (1714), is that the pursuit of self-interest by economic agents is beneficial to the economy as a whole and to the community. This belief is the foundation of liberalism and neoclassical economics. The second concerns the intrinsically beneficial character of technical innovations, which it would be advisable to encourage, or even to free up as much as possible, by relying on other future innovations to offset the ecological dangers they may entail. Finally, practical wisdom is a requirement to anchor action in concrete issues and observations. However, one of today's morbid absurdities is that of abstract indicators used to guide action without understanding what they produce.

Gross domestic product (GDP) alone illustrates all these criticisms that can be made of industrial modernity[11]. Created to evaluate the consequences of the 1929 crisis, it suited this question at that time (Kuznets 1934). Then it became a fetish indicator for economic policies, despite its limitations. Not only is it a poor indicator of what should be the main objective of public policies, well-being, but its growth comes at the cost of considerable social and ecological damage (Gadrey and Jany-Catrice 2007; Laurent 2019). The long-standing criticisms of this indicator have led to the development of alternative composite indicators that also take into account health, education, social links, etc. But the growth it measures remains the Holy Grail of politics. Yet, despite considerable effort, neither economists nor policymakers are able to grasp what makes growth possible, making their quest illusory (Banerjee and Duflo 2019). The continued importance of GDP illustrates both a drift in the economies of industrialized countries and an excessive reliance on numbers, no matter how irrelevant (Porter 1995).

Apart from environmental issues, there are also many signs that modern societies pay little attention to practical wisdom. In a work on the history of professionalism, I have shown that at the end of the 19th-century representatives of the professions most confronted with uncertainties (medicine and social work) were very aware of these uncertainties (Champy 2018b). Then with scientific progress

11 These criticisms are aimed more at financial capitalism than at family capitalism, which was more concerned with the solidity of companies, and therefore with the long term, and even with consumption, as in the Fordist model. It was also more anchored in concrete and known local realities.

(impressive in medicine in particular), the uncertainties that remained, however, were less and less assumed, and the imperative of practical wisdom became increasingly difficult to assert. Professional rhetoric, then, accords a far greater importance to objectivity, predictability and, much later, performance, which are impossible to fully satisfy in a situation of irreducible uncertainties, as the concept of practical wisdom helps to understand. The debates on food safety (Champy 2018c) or financial crises would provide other illustrations of the decline of practical wisdom in the epistemic sense. The latter is certainly not disappearing. Actors faced with uncertainties and aware of possible dangers are showing practical wisdom, and some are even struggling to restore frameworks of action more conducive to prudential practice (Champy 2019). But those who try to restore practical wisdom to the place that uncertainties justify are often hampered by the misunderstanding created when taking uncertainties into account, in social worlds where certainties and numbers should be the basis of all action. The history of the precautionary principle provides an illustration of the wall against which the demands for practical wisdom are met.

This principle is a prudential principle: it aims to prevent public actors from using the absence of scientific certainty to justify inaction in the event of a threat of serious and irreversible damage to the environment (Champy and Lepiller 2016). Debated in the 1980s, it was enshrined in European law in 1992 with the Maastricht Treaty, then transcribed into national law (in 1995 in France), and even enshrined in the French Constitution in 2005. This exceptionally rapid success is due to the fact that the principle responded to fears provoked by multiple crises. But it is partly illusory. From the 2000s onwards, criticism of the precautionary principle increased: notably, opponents of any new constraints on economic activity accused it of being an obstacle to innovation and growth. These arguments are gaining a growing audience, which is difficult to understand without taking into account an epistemological gap between this principle as a prudential principle, and a narrow conception of rationality for which only certainties can justify consequent efforts, the opponents of the principle considering it unreasonable to take costly measures in the name of possibly non-existent dangers. Thus, it is the very meaning of practical wisdom that they reject, as if rationality required that we only protect ourselves from dangers of which we are certain. Very recently, the proposal to include a principle of innovation in European law, to counterbalance the precautionary principle, illustrates the strength of reactions to the latter, in the name of freedom. The fate of the precautionary principle thus sheds light not only on the fragility of even moderate successes in the fight against environmental damage, but also on the difficulty of making prudential approaches understood and accepted. The rejection of practical wisdom as a logic of action is at the heart of industrial modernity.

9.4. Practical wisdom in the 21st century

9.4.1. *Reasons to doubt the relevance of this concept*

However, once a link has been established between a lack of practical wisdom in the past and the state of the planet, one might doubt that the issues of practical wisdom are still the same today. The first reason to doubt the topicality of this concept concerns the progress of reflexive attitudes: with the ecological, sanitary and financial crises, etc., we would have entered a period of greater reflexivity. I will not dwell on the answer to this argument, because all of the above shows the opposite. In particular, the myth of growth and the myth that technological innovations will solve all our problems remain powerful.

A more serious criticism concerns the usefulness of the concept of practical wisdom in actually guiding action in response to current threats. Three qualities of prudential action may cast doubt on the value of this logic. The first is its domain of relevance, which refers to situations of irreducible uncertainty. It is now established that disasters are already here or imminent. In each new report, the IPCC has been more assertive, removing all uncertainty about the reality of climate change, its anthropogenic origin and the fact that the human damage will be considerable, especially for the poorest (IPCC 2014). Chernobyl and Fukushima have made the risk of nuclear accidents tangible. The impasse to which the production of industrial waste leads and the depletion of non-renewable energy resources are just as certain. The decline of biodiversity is already visible. To face these dangers is only a question of information and intellectual honesty. Hence, the first criticism that can be made of the concept of practical wisdom: its relevance would weaken with the decrease in uncertainty about the reality of the dangers.

Secondly, practical wisdom often implies finding a balance between heterogeneous logics of action. For someone acting prudentially, the environment should not be allowed to overshadow everything else. Yet economic “realism”, or more precisely the obsession with growth and employment, has often been used to justify inaction, thus contributing to the catastrophe. By invoking practical wisdom, do we not risk reproducing this type of arbitration in which the environment always loses out? Hence the second criticism: the more the environment deteriorates, the more the order of priorities would be imposed on everyone, abolishing the need, inherent in prudential attitudes, to remain open to a modification of the hierarchy of the ends of action.

The third argument is close to the previous one. Well-anchored in reality, prudential action integrates the limits that circumstances impose on it. Deliberation on the right balance is consubstantial to it, which often moderates it[12]. But in the face of environmental urgency, does this search for measure not risk reinforcing the attitudes of denial of the extent of environmental threats? Hence the third criticism: responding to environmental degradation at its level would require a less pusillanimous framework of thought than that attached to the use of the concept of practical wisdom.

Of course, sometimes practical wisdom also requires audacity. However, the state of emergency of the planet no longer seems to fall within the type of situations where practical wisdom must be laden with audacity, but rather those where we should go for maximum audacity, without letting considerations of practical wisdom slow down action. We know what we need to do to fight global warming: consume less, increase the energy efficiency of our production and relocate it close to where we consume, substitute renewable energies for fossil fuels, and sequester CO_2 by planting trees (IPCC 2014; Jouzel and Denis 2020), all as quickly as possible. The boldness made necessary by the urgency thus seems to push practical wisdom out of the equation: the latter was useful in other historical contexts, but would now be at best obsolete and at worst counterproductive.

9.4.2. *The topicality of practical wisdom*

In contrast to this thesis, I will defend the idea that the gravity of the situation and its urgency reinforce the need for practical wisdom. For practical wisdom to become useless, all uncertainty would have to be removed. But the sources of uncertainty remain multiple. First of all, if the seriousness of several processes is no longer in doubt, these unfolding catastrophes do not exclude the appearance of still unknown dangers. Similarly, the precise consequences of climate change remain largely unpredictable: the global catastrophe that it constitutes comes in the form of a multitude of local disasters (due to cyclones, droughts, heat waves) for which mankind should be prepared without any precise certainty about the future. The debates about flood forecasting models and their uses show that what should be done is not completely self-evident (see Chapter 7). The impression of knowledge about disasters masks epistemic uncertainties that make precise knowledge of future problems illusory (see Chapter 8).

A second source of difficulty concerns action when these disasters are already here. Wars, industrial disasters or epidemics show how difficult it is to find the right actions, even when these disasters are visible to all and therefore known. Disaster

12 This is a criticism that has been made of the precautionary principle, not by liberal opponents, but, for example, by catastrophism theorists, such as Jean-Pierre Dupuy (2002).

creates chaos, a new world that is partly unknown, that puts habits and routines to the test, without it being obvious what can be done in their place. It is therefore prudential to question the habits and beliefs that were guides for life in order to find other guides. But these new guides are often difficult to find. Facing an emergency often requires the ability to prioritize and to take risks that would have been avoided in other times, as illustrated by disaster medicine. This chapter was written before the COVID-19 pandemic. However, the difficulties encountered by governments in deciding what actions to take in response to the pandemic illustrate the content of this paragraph particularly well.

Finally, I would like to show that, despite the certainties, practical wisdom is also required to try to slow down and limit climate change. Climate science tells us what should be done to have a serious chance of limiting the rise in average global temperature to 2°C between the pre-industrial period and the end of the 21st century. But the real question now concerns the means to achieve adequate changes in the economy, which requires diplomatic, regulatory, fiscal, legal and educational, actions, among others, involving international bodies, States, local authorities, companies, citizen-consumers, activist organizations, etc. The complexity of the task should encourage modesty and practical wisdom, without sacrificing the ambition and audacity that the urgency requires.

Criticism of industrial modernity is certainly progressing, but without always producing in response an increase in prudential reflexivity. Without claiming to deal with the concrete actions to be taken to fight against global warming, I will now try to show what a prudential attitude can consist of, relying on three dimensions of practical wisdom: its attention to the concrete consequences of action, mistrust of polarized beliefs and its relation to rationality and rationalism.

9.5. The imperatives of practical wisdom

9.5.1. *Practical wisdom seen as a consequentialism*

Practical wisdom is an invitation to judge the relevance of an action, in particular in relation to the consequences that can reasonably be expected[13]. However, the uncertainties and complexity I have just mentioned contribute to making the consequences of actions taken in the name of the environment uncertain: they do not always reach their goal. The environment is a field of action in which the gap between identified dangers, stated objectives of action and results actually obtained

13 Other moral considerations, for example, deontic, are not abolished, however.

is abysmal. A current requirement of practical wisdom, in environmental matters, is therefore a strong critical reflexivity to try to bridge this gap. In Weberian terms, the urgency and gravity of the stakes invite us to evaluate the relevance of an action more from the point of view of the ethics of responsibility than from that of the ethics of conviction. Wanting to do the right thing, to fight a noble battle, sometimes serves no purpose. Pragmatism must take precedence over purity of intention, and any discourse on the environment should lend itself to this consequentialist requirement. Let us take two examples that are sometimes combined: catastrophism inspired by the philosophy of Hans Jonas (1979) and the injunction to radical sobriety.

Catastrophism may have played a role in raising ecological awareness. One of its main theorists in France, Jean-Pierre Dupuy, has thus put forward the idea that announcing the catastrophe should make it possible to avoid it (Dupuy 2002). But this argument no longer seems valid today, as the discourse of collapsologists is gaining in popularity. The first criticism that can be leveled at it is that, since it has been held, this catastrophist discourse has not produced the expected practical effects. It even seems to be counterproductive, provoking effects of flabbergasting and discouragement, and even cynicism (Sapiains *et al.* 2016): What is the point of making efforts if the catastrophe is certain? Collapsologists themselves are abandoning Dupuy's perspective. For example, Pablo Servigne and Raphaël Stevens, authors of a book that helped popularize collapsology in France (Servigne and Stevens 2015), have since published another book whose purpose is to help prepare for the collapse of our civilization and for the world after (Servigne *et al.* 2018). Yet, from the perspective of action, it is better to consider that catastrophe is not certain, that it is avoidable, and that the outcome depends on the conjunction of public policies at all levels and countless individual decisions. The discourse on disaster closes this perspective rather than opens it, thus risking contributing to the acceleration of the disaster it announces. Yet disasters can be delayed and mitigated (and lives saved) if vigorous action is taken. The alternative that is needed seems to be immediate radical sobriety.

Sobriety is indeed an indispensable way to fight against global warming. But the consequentialist requirement also applies to it: Is asking consumers to radically change their behavior effective? Here again, history shows the opposite, since informed consumers who are aware of environmental issues continue to consume more and more. Growth certainly causes damage to the environment and to the quality of life in the long term, but it is also an addiction on which well-being depends in the short term. A detour through the economy of well-being helps to grasp the pitfalls of injunctions to swallow bitter pills (Sénik 2014). To announce degrowth and a radical change of lifestyle is to announce both a terrifying leap into the unknown for those who find in consumption a meaning lacking elsewhere, and the end of prospects for improvement that contribute to well-being in the present[14]. In order to imagine a way out of this dilemma in a democracy, one has to assume that consumer preferences can be changed (Banerjee and Duflo 2019), that is, that consumers can accept that their well-being no longer depends on consumption to the same extent, or in the same way. A prudential attitude, that is, one that is both grounded in the difficulties of change and concerned about its concrete consequences, should consist of seeking ways to reorder these preferences and thus create the conditions for sobriety[15].

The task is hard enough that several levers should be used at the same time. The first is of course education and information, so that everyone is as aware as possible of the precise ecological consequences of their consumption (travel, meat, heating, manufactured products, Internet use). But actions depend less on knowledge and convictions than on previous actions. On this basis, the theory of commitment explains how to obtain from a person an action that they do not desire *a priori* (Kiesler 1971; Joule and Beauvois 1998). Getting people to perform small, seemingly insignificant actions greatly increases the chances that they will agree to perform more costly ones that they associate with the previous ones[16]. This scheme is widely used in areas such as public health and in the fight against energy waste, and could be used to provoke commitments to ecological objectives that would thus be socially valued and could become increasingly ambitious. Second, well-being depends strongly on interpersonal comparisons, through the ostentatious dimension of consumption. However, the mobilization of young people in particular can help to promote new norms and to create an emulation that no longer supports this conspicuous consumption, but rather the conspicuous orientation toward consumption that is less costly in terms of energy and produces less waste. These norms already exist in certain circles and all the levers for their diffusion should be

14 These two seemingly similar challenges actually refer to two distinct factors of well-being: adaptation and a taste for progression (Sénik 2014).

15 Luc Sémal invokes community experiences of sober living to answer criticisms of the ineffectiveness of catastrophism (Sémal 2019). But he neglects the following fact: although insufficient, a 10% decrease in greenhouse gas emissions by 90% of the population would be a hundred times better than a 90% decrease by 0.1% of militant pioneers. The only hope is to change the lifestyle of as many people as possible, and to reduce inequalities because the richest consume the most. Radical initiatives are useful when they are inspiring, but insufficient in themselves, and even harmful if they act as a repellent.

16 The means of this powerful commitment, which I will not present here, are the exact opposite of guilt-tripping.

used, following the example of environmental referents in the organizations (as they are equipped with referents for gender equality)[17].

My point is not to say that small reductions in energy consumption will be sufficient, or that everything should come from consumption, even if the reduction of the latter is unavoidable. It is not even to deal with a question as complex as the evolution of consumption. It is only to highlight the differences between prudential logic and an approach that sets out goals without dealing with the concrete means of achieving them, at the risk of producing the opposite. As much as possible must be done to act immediately, which implies thinking about the conditions for ambitious actions that are effective in the social world as it is, and not only about ideal actions, intellectually attractive but sometimes counterproductive. Consumers do not convert to making major sacrifices, and injunctions to wean themselves off overconsumption can be counterproductive. It is therefore essential to start from what economics and psychology teach us about how their well-being is constructed in order to change their preferences and make possible freely consented reorientations of consumption, as part of a new "social contract" on climate (Caminel *et al.* 2014, p. 71).

9.5.2. *Practical wisdom versus polarized beliefs*

One of the preconditions mentioned above is not to give in to the temptations of false debates (Angenot 2008). The theme of degrowth is so successful because it responds to the myth of growth. But we have just seen that presentations of the theme of degrowth are too unattractive to be an immediately effective solution. A team of economists who cannot be suspected of taking environmental issues lightly have shown that the need to lower our CO_2 emissions is such that changing our consumption will not be enough, unless it is carried out so brutally that it could not be a democratic choice (Caminel *et al.* 2014). Other solutions are required, which leads us to address the issue of technical progress.

One belief of industrial modernity is that the solutions to our problems will come from technical innovations. But it is contradictory to expect technical–industrial development to provide solutions to the environmental degradation it has caused, without having been subjected to criticism and reform. Thus, apparently attractive technical solutions often cause pollution to be displaced. Engineers' claims about the energy efficiency of new techniques have often been disproved by the facts. Let us

17 Another lever is a tax incentive, through the introduction of a carbon tax. But it must be made acceptable by compensating those with the lowest incomes, which illustrates the prudential requirement to understand problems in their entirety, and not on the basis of technical divisions that deny their complexity.

look at a powerful spring of the growth myth: the belief in decoupling. The idea is that growth will continue without the energy costs and damage it has caused so far. More precisely, three decouplings of growth are announced: from greenhouse gas emissions, energy consumption and other material consumption[18]. But not only have these decouplings not yet taken place, or only very punctually in time and space, but the first two will be increasingly difficult to achieve. Indeed, the amount of energy needed to produce a given quantity of energy is increasing sharply, whether it be oil extraction or the production of equipment to harness renewable energy. This is the theme of "energy cannibalism" (Pearce 2008), which can be applied to energy transition. For example, above a certain level of growth of a solar panel park, the production of new panels would absorb all the energy supplied by the existing panels. Thus, one of the limits of the energy transition is that it is itself energy-intensive, which makes it unreasonable to expect technical progress alone to solve the problems of the Anthropocene[19]. Finally, the rare progress sometimes obtained has been cancelled out by a rebound effect, that is, by the increased consumption that it has itself encouraged: for example, when a car consumes less, it becomes tempting to travel more[20].

The obstacles to decoupling make it essential to reduce consumption in order to mitigate the ills of the Anthropocene as much as possible. But, given the difficulties of reducing consumption, it would be unreasonable to give up all technical innovation. The past offers some encouragement, such as the decoupling following the oil shocks of the 1970s, even if it was short-lived. Without seeking to know the extent of possible decoupling, and without pinning all our hopes on it, we should make it a political priority, for example, by providing massive incentives for housing insulation and by providing more funding for research into energy storage[21]. Finally, even if it is limited, decoupling can contribute to the commitment in concrete, immediate and beneficial actions of consumers who oppose the idea of degrowth.

Decoupling is one of the means available, provided that we are aware of the pitfalls it entails: excessive confidence in its contribution to the fight against climate change, rebound effect and the risk that local decoupling, made possible by the relocation of productive activities to countries in the Global South, will mask the

18 See Caminel *et al.* (2014, p. 18). I rely entirely on this work for this passage, without being able to summarize it as accurately as it deserves.

19 A practical consequence stated by Pearce is that technologies with the fastest energy payback should be deployed first, and where they are most cost-effective.

20 Unlike other difficulties, the rebound effect could be combated by means such as consumer education and tax incentives.

21 Production and consumption of renewable energies are often misaligned in time, which makes their use dependent on their storage.

overall increase in damage. We should not want to choose between degrowth and decoupling, which are often opposed in public debates: both can help and neither will suffice on its own. One of the difficulties of a prudential attitude is that it implies avoiding polarized beliefs that structure the public debate at the expense of the required complexity.

9.5.3. *The procedural requirements of rationality*

Finally, the concept of practical wisdom allows us to reconsider the question of rationality. In circles sensitive to environmental issues, a negative value is often ascribed to "rationalization", as it is associated with the triumph of economic reason and polluting agricultural and industrial techniques. However, if the process of rationalization designates a progression of rationality, it should be recognized as having a positive value: Is not rationality better than irrationality for any major decision? The rejection of the idea of rationalization can be explained by its progressive appropriation, during the 20th century, by the supporters of a scientistic and computational conception of rationality, which shows its limits in situations requiring practical wisdom. To accept the progress of rationalization means to accept the domination of economic reason, the authority of scientists and experts who have often minimized risks, and the managerial criteria for evaluating activities and their performance. Changes in organizations illustrate this last point (Champy 2018b), as does GDP as an indicator of growth (see above). But rationality is not what the proponents of this "rationalization" have made it. It is, on the contrary, an achievement of thought built up over the centuries since Greek antiquity, whose essential dimensions are, as with practical wisdom, attention to reality, a strong reflexivity and a capacity for doubt (Petit 1995). In a situation of irreducible uncertainty, practical wisdom constitutes the only possible ethical rationality. Rediscovering the concept of practical wisdom thus makes it possible to challenge the discourse on rationality of the proponents of "rationalization", without falling into irrationalism.

The concept of practical wisdom then helps to clarify the social conditions of rationality in uncertain situations. A crucial element of this clarification concerns the procedural dimension of rationality. Indeed, substantial rationality is inaccessible in case of uncertainty, since it would suppose an incontestable substantial criterion (often scientific knowledge) of what is rational[22]. When this criterion is lacking,

22 The economist Herbert Simon theorized an opposition between substantive and procedural (or limited) rationalities, which also starts from the observation that it is impossible to achieve substantive rationality, which would presuppose an exhaustive knowledge of the world (Simon 1982). Simon made the search for a satisfactory decision (as opposed to an impossible optimal

means must be found to reduce the risk of harmful errors. Practical wisdom indicates several complementary means. One of them consists of sticking to provisional and reversible actions, the time to increase the available knowledge. Another is to rely on the judgment of those best prepared, if possible by collecting several views. In the face of uncertainty, the search for rationality must therefore be based on collective intelligence: practical wisdom can no longer be seen as an individual virtue. This way of substituting the objectivity of science, whose limits are encountered, with the intersubjectivity of people whose judgment is sharpened by experience, is used in several fields of activity: for instance, in economics with panels of experts, and in medicine with consensus conferences. Not only does it bring together complementary competences, but it also allows divergent points of view to be expressed in order to bring about a clearer vision than the initial individual visions. But for these arenas to fulfill their role, one must be careful to avoid the possible effects of a common cognitive bias, conformism within the group, or even conflicts of interest. This is why this collective intelligence must be organized by procedural rules: systematic declaration of interests, rules of collective deliberation allowing everyone to speak up, protection of people expressing a divergent viewpoint, diversification of the experiences on which participants can draw, sufficient time for in-depth discussions aimed at making explicit the presuppositions on which each position is based, etc.

An interest in the relationship between rationality and practical wisdom also leads us to reconsider the place to be given to numbers and scientific knowledge in deliberations on the reasons for action. Excessive confidence in numbers is a flaw in contemporary rationalism. But criticism of quantification sometimes takes the form of a general disqualification, including in scientific circles. It is easy to show that without quantitative notions, climate change would have gone on unnoticed. Figures and data are thus needed to steer public action, to evaluate it or to negotiate shared objectives in climate conferences. It is therefore necessary to consider the criticisms made of them. A recurrent criticism is that they are schematizations of reality. Yet, every speech is a schematization of reality. The real challenge is to establish a critical relationship with figures, which allows them to be used with practical wisdom. It is a matter of choosing indicators that are adapted to the issues that we wish to address, knowing that any measure is underpinned by an objective whose political value must be questioned; of multiplying them to provide a sufficiently rich

decision) the object of procedural rationality. But the concept of practical wisdom differs from that of satisfaction. First, it helps us to think that what seems satisfactory may carry great dangers. Moreover, in some situations, no truly satisfactory solution is available. The decision may then aim at minimizing the risks taken.

image of the complex realities that are the focus of action; of not making them say anything other than what they can say, etc.[23].

Practical wisdom implies refusing the ease of simplifying decision-making by clinging to false certainties or oversimplifications of reality. But the concrete social conditions for avoiding these errors can be improved. Escaping the polarization between irrationalism and narrow rationality (managerial, scientistic and technicist) gives us the means to reflect on the procedural conditions of practical wisdom, and thus on the ways of another rationalization.

9.6. Conclusion

Science took as its object of reflection the natural world, which was external to it. Then, critics of modernity invited us to bring critical reflexivity to focus on scientific work, on its products and on their uses (Beck 1992). From now on, the condition for more practical wisdom is that this critical reflexivity also concerns the actions undertaken and the discourse held in the name of the environment. This is not specific to the present historical moment: in all eras, practical wisdom implies a strong attention to the consequences of actions. Yet emphasizing this point makes sense. When the first observations and findings led us to talk about the Anthropocene, the main challenge of a prudential attitude was to pay attention to even weak and uncertain signs of danger, and to take them seriously despite the uncertainty. Being open to questions like "Has the Earth entered a period of global warming? If so, is it anthropogenic?" was in itself prudential when the IPCC began its work just over 30 years ago. Refining the answers to these questions was already a contribution to the fight against (possible) anthropogenic global warming. This has not been enough for a long time: research and observation remain useful, but action is urgent. Yet the world is so complex that knowing what measures should *ideally* be taken is no longer enough. It has even become a small part of the problem. The necessary changes (in consumption and production patterns, for example) are well known, but their rapid implementation faces obstacles. The path that would allow us to avoid these obstacles has thus become more decisive than the objectives themselves, which nevertheless continue to focus public debate.

Let us consider it a given that practical wisdom is more necessary than ever. But can the *concept* of practical wisdom help in any way? Unearthing it today might seem to imply an exaggerated effort for uncertain benefits: this concept has become difficult to grasp since the 18th century, when technical rationality imposed its

23 For a detailed commentary on the qualities of a good indicator, see Laurent (2019). However, the concept of practical wisdom would allow us to go a little further than this author does.

domination (Delannoi 1995), and knowing it does not replace the skill and judgment needed to find an answer to the question: "What can we do?" But we have just seen that this philosophical knowledge can help to prevent gross errors. It draws attention to the pitfalls to be avoided to satisfy the requirements of practical wisdom and prevent damage: indifference to the concrete consequences of actions, often with the best of justifications (notably purity of intent), polarization of beliefs (reinforced by understandable opposition to questionable contrary positions) and the rejection of rationality and the tools it provides (science and numbers, often misused).

The concept of practical wisdom thus also makes it possible to reclaim the ideas of rationality and rationalization, which have been confiscated by actors who have a narrow and, above all, fallacious conception of them. It thus helps to grasp that the impasse of irrationalism is not the only alternative to this conception. Using ancient philosophical reflections on practical wisdom then gives practical help in two ways: to grasp which epistemic and deontic attitudes are appropriate to the uncertainties and abysmal dangers of the Anthropocene; to make explicit the social conditions that are conducive to making these attitudes less difficult to assert. Knowledge of the concept of practical wisdom is thus a resource for increasing the effectiveness of the actions to be undertaken to restore spaces for deliberation and prudential debate[24]. A prerequisite is then to grasp that the same difficulties run through fields as different as healthcare (Champy 2018b), social work (Kuehni 2019), or even economic policies and the fight against global warming. Conversely, the false debates of which I have presented examples show that it would probably be illusory to want to break with the shortcomings of industrial modernity without having clearly identified what is probably the most important of them: the denial of the imperatives of practical wisdom. The real audacity is there: to attack the heart of ways of thinking, in order to produce another relationship with the world. In terms of environmental emergencies, another requirement is added: facing up to the uncertainties and limits of knowledge must not hinder the necessary determination. These reflections outline a horizon of strong requirements for reflexivity and action at the time of the Anthropocene.

9.7. References

Angenot, M. (2008). *Dialogues de sourds : traité de rhétorique antilogique*. Éditions des Mille et Une Nuits, Paris.

Aubenque, P. (1963). *La prudence chez Aristote*. PUF, Paris.

24 Let me refer to a method for attempting to reduce barriers to practical wisdom in healthcare in bureaucratic organizations, which I have presented elsewhere (Champy 2019).

Banerjee, A.V. and Duflo, E. (2019). *Good Economics for Hard Times.* PublicAffairs, New York.

Beck, U. (1992). *Risk Society. Towards a New Modernity.* Sage, London.

Broadie, S. (1991). *Ethics with Aristotle.* Oxford University Press, New York.

Caminel, T., Frémeaux, P., Giraud, G., Lalucq, A., Roman P. (2014). *Produire plus, polluer moins : l'impossible découplage ?* Les Petits Matins et Institut Veblen, Paris.

Champy, F. (2015). Pourquoi le soin n'est pas qu'une question de technique. In *Soigner l'humain. Manifeste pour un juste soin au juste coût*, Georges, C., Astre, H., Pierru, F. (eds). Presses de l'EHESP, Rennes.

Champy, F. (2018a). The sociology of prudential actvities. From collective commitment to social innovations. *Sociologia Problemas e Práticas*, 88, 79–94.

Champy, F. (2018b). What are the issues of focusing on irreducible uncertainties in professional work? A historical outline of "prudential professionalism". *Cambio. Rivista sulle Trasformazioni Sociali*, 8(16), 15–30.

Champy, F. (2018c). Food safety and consumer rationality. Is a "food pact" possible? In *Food Safety in Japan and China. Theoretical Perspectives and Empirical Insights*, Augustin, L., Poulain, J.-P. (eds). Routledge, New York.

Champy, F. (2019). Le travail de soin et les mobilisations éthiques en faveur de la "sagesse pratique". Des réflexivités réformatrices encastrées dans des bureaucraties. In *Trajectoires d'innovation. Des émergences à la reconnaissance*, Klein, J.-L., Boucher, J.L., Camus, A., Champagne, C., Noiseux, Y. (eds). Presses de l'Université du Québec, Quebec.

Champy, F. and Lepiller, O. (2016). Le principe de précaution. Robustesse pragmatique et fragilité sociale. In *Faire face aux risques dans les sociétés contemporaines*, Becerra, S., Lalanne, M., Weisbein, J. (eds). Octarès, Toulouse.

Delannoi, G. (1995). La prudence dans l'histoire de la pensée. *Mots*, 44, 101–105.

Dupuy, J.-P. (2002). *Pour un catastrophisme éclairé.* Le Seuil, Paris.

Gadrey, J. and Jany-Catrice, F. (2007). *Les nouveaux indicateurs de richesse.* La Découverte, Paris.

IPCC (2014). AR5 Climate Change 2014. Synthesis report, Intergovernmental Panel on Climate Change.

Jonas, H. (1984). *The Imperative of Responsibility: In Search of an Ethics for the Technological Age.* Chicago University Press, Chicago.

Joule, R.-V. and Beauvois, J.-L. (1998). *La soumission librement consentie.* Paris, PUF.

Jouzel, J. and Denis, B. (2020). *Climat, parlons vrai.* François Bourin, Paris.

Kagan, D. (1998). *Pericles of Athens and the Birth of Democracy.* Free Press, Glencoe.

Kiesler C. A. (1971). *The Psychology of Commitment.* Academic Press, New York.

Knight, F. (1921). *Risk, Uncertainty and Profit.* University of Illinois, Champaign, IL.

Kuehni, M. (ed.) (2019). *Le Travail social sous l'œil de la prudence.* Éditions Schwabe, Bâle.

Kuznets, S. (ed.) (1934). National Income: 1929–1932. *National Bureau of Economic Research Bulletin*, 49 [Online]. Available at: www.nber.org/chapters/c2258.pdf [Accessed 22 May 2020].

Laurent, E. (2019). *Sortir de la croissance, mode d'emploi.* Les liens qui libèrent, Paris.

Oreskes, N. and Conway, E.M. (2010). *Merchants of Doubt.* Bloombury Press, London.

Pearce, J. (2008). Limitations of greenhouse gas mitigation technologies set by rapid growth and energy cannibalism. *Klima 2008.* DOI:10.31219/osf.io/wjst5.

Petit, A. (1995). La rationalité. In *Notions de philosophie III*, Kambouchner, D. (ed.). Folio, Paris.

Porter, T.M. (1995). *Trust in Numbers: The Pursuit of Objectivity in Science and Public Life.* Princeton University Press, Princeton, NJ.

Sapiains, R., Beeton, R., Walker, I.A. (2016). Individual responses to climate change: Framing effects on pro environmental behaviors. *Journal of Applied Social Psychology*, 46(8), 483–493.

Semal, L. (2019). *Face à l'effondrement. Militer à l'ombre des catastrophes.* PUF, Paris.

Sénik, C. (2014). *Économie du bien-être.* Le Seuil et La République des idées, Paris.

Servigne, P. and Stevens, R. (2015). *Comment tout peut s'effondrer.* Le Seuil, Paris.

Servigne, P., Stevens, R., Chapelle, G. (2015). *Une autre fin du monde est possible.* Le Seuil, Paris.

Simon, H.A. (1982). *Models of Bounded Rationality.* MIT Press, Cambridge.

Conclusion

Julien REBOTIER
TREE, CNRS – E2S UPPA, Bayonne, France

This book has taken stock of what the notion of the Anthropocene can bring to a rich existing scientific literature on risk, especially in geography. With its relatively recent but intense and continuing popularity, the Anthropocene invites us to reinterpret environmental issues, including risk.

C.1. Risk and the Anthropocene, a problematic connection

However, as this book has shown, this rapprochement poses several problems. We will develop two of them for the purposes of this concluding statement.

First of all, the notion of the Anthropocene intervenes mostly in parallel rather than as a complement to decades of accumulated knowledge from research and action programs on risk. It seems easier to place under a new emblem another layer of discourse that actually deals with the issue of risk than to evaluate existing productions, to take the measure of the recurrent obstacles on the way to a finer understanding and more virtuous action, or to reformulate the questions (for research as well as for action) in the light of these elements. The contributions in this book have shown how the "emergency" of positioning and acting on risks in the context of the Anthropocene should not lead us to neglect the contributions accumulated to date. The book's various chapters are relevant today and often point out the obstacles and blockages to which the transition to a narrative register of the Anthropocene does not change much.

Risks and the Anthropocene,
coordinated by Julien REBOTIER. © ISTE Ltd 2021.

Second, the notion of the Anthropocene can act as a powerful rallying banner, but one under which the expression of diversity is not given. It makes a common cause with global society and with risk confronted with an unprecedented, global extremity... and exclusionary in many respects. Yet the difficulty of thinking of the world in terms of the Anthropocene is obvious as soon as we try to adopt reasons for thinking and acting other than those that have presided over the introduction and success of the notion, of naturalist origin, on the international scene. Thus, the Anthropocene moment does not seem to make more room for power relations, asymmetries, and divergent interests between actors, scales and territories than do risk studies in their great diversity. The challenge remains unresolved, redoubled nonetheless by an urgent need to act. The critical and reflexive approach of the social sciences, which reintegrates the objects studied as well as their definition and study in their contexts, continues to find many justifications here.

It is not a question of giving up on a better understanding of the biogeochemical mechanisms involved in natural hazards, nor those of the climate and their evolution. But from the point of view of the social sciences, risks such as the Anthropocene are linked to social issues (specific to social relations and to relations between societies and their environments) that are posed by the environment. In this respect, the environment (the problems it poses, which have a history) is a social issue. The climate cannot be the cardinal object of research mobilized to respond to the challenge of climate change. As for the sciences of the earth system, do they have the vocation to conceive and direct the research from which the social world is invited to think? It is up to the social sciences to define the questions, the objects and the scope of their research. The Anthropocene, even if it is deconstructed and criticized by the social sciences, seems to function as a performative category (which tends to impose a conception of the world), more than as a heuristic category (which allows meanings to emerge around conceptions and relationships to the world). However, one can consider that the political harmlessness of the frameworks of thought is an important condition of their success, as is their capacity to consolidate what already exists. History stutters more than it repeats itself. We never witness quite the same, and things, even minor ones, change. If the domination at work remains powerful, in the scientific field as much as in people's minds, what common and plural future can the Anthropocene be called?

C.2. Relationships to be built in order to draw up research perspectives

In addition to questions about the consequences of climate change on extremes or the frequency of natural events (potentially associated with major disasters), the relationships between risks and the Anthropocene need to be constructed, in particular on the basis of specific contributions from the social sciences. It is appropriate to consider the possibilities of providing the social sciences with a more self-sufficient research agenda on risks in the midst of the Anthropocene.

What does the notion of the Anthropocene bring to scientific production on risks? How do the social sciences, with their critical apparatus, deal with this omnipresent notion today? How do the social sciences position themselves with regard to risks and the environment, at a time when emergency seems to be the order of the day? Does the rise of the Anthropocene correspond more broadly to a particular moment, understood as a historical period characterized by ways of thinking, acting and doing? What if the Anthropocene moment were an opportunity to anchor risk research in society, alongside and in connection with the earth system sciences, rather than in parallel with or downstream from perimeters that the latter define as appropriate?

It is this range of problems that the book has set out to explore in three parts presented in the form of questions. For each of them, three chapters have provided some answers, from which some synthetic lessons are gathered here.

In the first part, "Toward unprecedented risks?" Chapters 1–3 showed how, beyond the Anthropocene, risks are permanent and sensitive to spatial, social and historical contexts. The rise of the Anthropocene question, nevertheless, allows us to think about scalar articulations, while the scale at which a risk is posed and defined contributes to configuring this risk. Thus, without really considering new risks, the Anthropocene brings certain scales, territories or themes to the fore (as can be the urban climate). We are then faced not so much with the new risks introduced by globalization and identified at the end of the 20th century, highlighting certain aspects of risk to the detriment of others.

The uneven emphasis on risk in the Anthropocene moment should provoke critique and reflexivity among the social sciences.

In the second part, "Recompositions for the study and management of risks?" Chapters 4–6 tend to answer in the negative. Here, too, great permanency is described, whether it is a question of the instrumental and political dimension of environmental restoration measures, or whether one is confronted with the permanent difficulty, over the course of the last decades of the 20th century, of moving away from the naturalistic explanation to bring the cause of risks and disasters to an international level. This part almost identifies the inertia of the categories of analysis, and the difficulty in making them evolve over time, despite the feedback allowed by the progress of knowledge. In the same way, the construction of international management policies and the social conditions for the production of knowledge on risks, before and during the Anthropocene, are very rigid, which contributes to confining risks within certain perimeters, questions and horizons of possible solutions, from which it is difficult to escape. With the transition to the Anthropocene, it seems easier to change words than to ask other, more sensitive questions about the contestation of power relationships, the distribution of efforts to be made, or political choices about desirable futures to be debated.

More than the notion itself, the Anthropocene moment allows us to take a position on the meaning of environmental research and the effects it has on the world.

In the last part, "What consequences for a changing modernity?" Chapters 7–9 review the already documented transformations of a modernity for which impossible control and irreducible uncertainties are obvious. The shift toward population control and, more broadly, the generalization of managerial modalities shows how recent developments in risk analysis and management are an opportunity to socially redistribute the burdens of responsibilities, efforts, choices and consequences in space, in time, and within societies. The advent of liberal societies, in full retreat from the control of risks that have become diffuse, corresponds to a form of de-collectivization specific to a modern State in mutation. The declension of this trend in the different registers of State intervention (including in the field of risk) raises questions about the capacity to create society, to create a common ground, while the age of the Anthropocene paradoxically calls for "sacred union", for our shared humanity. The imperative of prudence presented last aims not to leave the privilege of rationality to managers while technical and scientific rationalism prevails in the face of the complexity of the world. It is advisable, modestly, through prudence, to establish a different relationship with the world that re-inscribes solidarity, among societies and with the rest of nature, as a principle of a shared world.

Thinking and recognizing the conditions of thought, the social sciences in environmental research contribute in this way to the emergence in practice of future worlds in their greatest diversity.

Each of the three parts of the book – paying attention to the differentiations of the world implied by the Anthropocene, adopting a permanent reflexive approach to the scope of social science research on the environment in the Anthropocene era and having the ambition to better understand as well as to accompany the contemporary world and possible futures – and their main teachings define the classical contours of a programmatic horizon for social science research on risk. The interest of this scientific horizon is redoubled by the Anthropocene moment, which underlines its relevance. It remains to be seen whether this moment will also bring about a significant difference in the production and assessment of risk studies that now cumulate several decades of work.

C.3. Taking on the dual challenge of the environment and the social role

It is therefore interesting to end the book by proposing four transversal reading axes, as a return to the different chapters and their links, bearing in mind the perspective of a research horizon for risk studies in the midst of the Anthropocene moment. These axes refer to two main aspects of social science research. The first concerns what it is about, about the environment, risks and the Anthropocene (the "about what" of research); the second concerns the way in which it is articulated to the rest of scientific practices, to utility and to the social world in general (the "how", or even the "why" of research).

C.3.1. *Writing at the time of COVID-19*

Most of the chapters in this book were written or worked on during the COVID-19 pandemic and the most stringent containment measures (first half of 2020). The episode is sometimes explicit, sometimes not, but resonates with the thoughts developed. Is COVID-19 a manifestation of the threats of the Anthropocene? A tendency for risks to become more complex, globalized and uncertain? Does the global sequence of COVID-19 herald a redefinition of the place of the state, the public good and the general interest? As an *in vivo* experiment of global disaster, COVID-19 shows how much damage varies depending on crisis management, but also according to the answers given (or not) to previously identified problems. The unity of the threat is quickly eroded, both in the forms it takes and in the damage it implies, in space and among societies. If this theme, though present in France[1], remains discreet, it is more widely covered in the English-speaking press in relation to caregivers from immigrant backgrounds (*New York Times*) or the "BAME" (*Black, Asian and Minority Ethnic – The Guardian*). Management policies are being challenged because of the paradox they illustrate. The global nature of new threats is increasingly being identified, while responses are becoming nationalized and fragmented, including within the European Union. The nation-state, a level that is

1 The results of the first part of the EpiCov survey (www.epicov.fr/) conducted by INSERM with the assistance of Drees, Insee and SpF were made public on October 9, 2020, just as this text was being written. They show how the pandemic (both the transmission of the disease and the management measures, such as containment) has accentuated social inequalities in a cumulative manner.

constantly being transformed in a recomposed modernity, is coming back as the main actor in health care, in place of transnational actors, multinational firms and other instruments of international governance. Can the world scale oblige other than through the motivations of economic globalization? Can the world compel in the sense that it imposes new solidarities that break with the ideology of progress or that of the market?

C.3.2. *The Anthropocene thought on a global scale, a force for homogeneity*

This is another possible pitfall of the Anthropocene that is frequently recalled in various chapters. It remains very difficult to articulate local and global dynamics without presuming one over the other, without bias of principle, nor abusive transfer of responsibilities. For many, the notion of our shared humanity obliterates the nuances and diversity of the local, sacrificed to the imperative "our common future" of the Anthropocene, as an echo to that of sustainable development. This shared humanity serves as a fable in the face of the infinity of small, narrowed, communalized "we", of class, of origin. The "we" of the *Song of the Workers*, in the introduction of this work, is historically marked. It exists at the same time because of a consciousness of oneself and to a collective movement, which the repressions of the revolutions of 1848 have annihilated in violence. Does the Anthropocene moment lead each of us to *choose* to be sensitive together and to react with a collective movement that makes (or will make) historical sense? The change of scale is also a change in the way of thinking about the relations between humans and about humans with (the rest of) nature. The emergence of the World, the scene of the Anthropocene, is translated into globalization (cultural, but also economic and financial), or even planetarization, presuming different basic relations. Here again, the triumph of individualization and of a fragmented world strangely contrasts with the recognition of a common destiny of humanity.

C.3.3. *Two diversionary effects of climate change and Anthropogenesis*

First of all, reinforcement of the naturalization implied by the Anthropocene is mentioned in many places. Our gaze is focused on the threats and diverted from the issues as well as from the genesis of environmental problems. The focus is on perceptions and representations on the gaps between the Anthropocene as described by the Earth sciences and the responses or ideas of social actors. It is the resulting vulnerability (from exposure to threats that are hardly questioned) rather than contextual vulnerability (the result of socio-historical relationships) that is highlighted. Second, the idea of a major environmental challenge that the

Anthropocene conveys obliterates other types of risks and problems linked to the social world (in which environmental challenges are nevertheless posed!). The climate, earth and life sciences cannot be the flagship of environmental research. Green thinking requires more than sticking to the arguments of the sciences that measure and that base their evidence in numbers. Numbers do not speak any better than speeches (at least, just as badly!). One can resort to an empathetic rationality (beyond the human race), whose driving forces are the refusal of domination and subjection and the respect of diversity. Despite what the radical *anthropos-* implies, here misleading, the Anthropocene does not necessarily build bridges within humanity on Earth. It can even opacify the asymmetries of the world. The environmental emergency and the thinking of the whole deny the less fortunate their very status as dominated and lock them into a double punishment: they are both the most vulnerable and those whose voice counts the least (or for whom we speak the most). At what point does the risk become unbearable? The need to act is more likely to urge because it concerns the world of the well-off than because it threatens the "web of life", when many other worlds have already collapsed. Without much effort racking the brain, Rachel Carson's *Silent Spring is* already more than half a century old. Moreover, the "moralization of capitalism" once evoked by a certain president of the French Republic did not withstand the recovery of the 2008 financial crisis. These questions are not technical, and they do not concern knowledge of the earth system or stratigraphy, nor the ability to reduce uncertainties. They concern deliberation, politics and the need to make choices that are binding, including the practices of science.

C.3.4. *Back to modernity*

Since the end of the 20th century, reflections on risk have been strongly marked by those on modernity, the universal and the recomposition of the State and societies. The emergence of reflexive modernity, however, hardly calls into question the foundations of modernity (progress, technology or science). The mobilization of the individual and the local is paradoxical, between the promotion of a liberal agenda (taking charge of oneself by oneself, transfer of responsibilities) and the demand for the autonomy of local actors, for contextualized and emancipating approaches. The modern project, that of individual and collective progress at the same time, that of freedom from material contingencies, has been partly achieved... but at what cost! Through the question of risks, the Anthropocene provides another opportunity to revisit a troubled modernity. The non-correspondence between the globalization of threats and the scales of risk governance inspires reflection on the respective roles of politics and economics. The expansion of the liberal field extends the front of appropriation that sees risks through probabilities and especially insurance. The securing of uncertain futures accentuates the technical and instrumental characteristics of modernity, even though they are denounced by the reflexive exercise. It does not mean less damage, but the possibility of subjecting futures (including damage) to a computational rationality, a form of management. This generalized translation requires a considerable improvement in the "capacity to monitor and manage risks on a large scale, in order to be able to act more quickly, as soon as the first signals appear"[2]. As the COVID-19 episode shows, the State responds as a last resort. As for the insurance logic, it tends to colonize the future through market logic. But it remains to be seen how the State will interact with it. In the long run, the state that continues to intervene will do so with what political aims? Far from the limits of the planet, do ethical limits – and aims – have a place in risk governance?

In the age of the Anthropocene, risk management is no longer just about reducing the damage that we know is inevitable. It has also (especially?) become the management of crises and permanent uncertainties, between biopolitics (introduced by Foucault) and technoscience (illustrated by Crutzen). The levers of action have been shifted to individuals who have become the primary artisans of the survival and destiny of humanity. We scrutinize their levels of acceptability and the right behaviors as a guarantee of their ability to adapt. The planetary threats and the radical uncertainties linked to them constitute eminently legitimate research areas – in climatology, ecology, virology, etc. They are reinforced by social demand and management expectations in the Anthropocene era.

However, these radical uncertainties should neither compromise nor condition our modes of action. On the one hand, there is no simple causal link between hazards and disasters. On the other hand, we already know a lot of elements of vulnerability that are the breeding ground for future disasters. The deleterious housing conditions and their over-occupation, the precarity of the means of subsistence and deficient social protection, the degradation of the health services and the lack of sanitary facilities, exposure to occupational risks and a differentiated state of health according to social class, gender or ethnic or migratory origin, etc., all these elements were known long before the health disaster[3] that we too often hear, and wrongly, that it was "caused by" COVID-19.

So what more do we need to know to act on risk? Or put in another way, why does not what we already know allow us to act? "Our house is burning down and we are blind to it", said Jacques Chirac at the Johannesburg Earth Summit in 2002. No doubt disaster studies benefit from not being blind to anything, from assuming as

2 Interview by Erwann Michel-Kerjan, *Le Monde*, April 24, 2020, p. 26.

3 For the French case, these elements are confirmed by the first results of the EpiCov survey mentioned above.

central to their research agendas elements of vulnerability – of development, of inequality, of domination – which *a priori* do not fall within the domain of risk in order to return even more effectively to that which causes disaster. An additional step is to do this while recognizing the universe of multiple constraints in which the study and management of risks evolve, if not to turn the tables, at least to avoid deluding ourselves when envisioning future worlds.

List of Authors

Christine BOUISSET
Université de Pau et des Pays de l'Adour
TREE
CNRS – E2S UPPA
Pau
France

Florent CHAMPY
LISST
CNRS
Toulouse
France

Julia HIDALGO
LISST
CNRS
Toulouse
France

Kristian KRIEGER
Institut ISOPOLE
Université catholique de Louvain
Belgium

Pascale METZGER
PRODIG
IRD
Paris
France

Catherine MEUR-FEREC
LETG
Université de Bretagne Occidentale
Brest
France

Sébastien NOBERT
Department of Geography
Université de Montréal
Canada

Florian PAPPENBERGER
ECMWF
Reading
United Kingdom

Patrick PIGEON
MÉDIATIONS
Université Savoie Mont Blanc
Chambéry
France

Julien REBOTIER
TREE – E2S UPPA
CNRS
Bayonne
France

Magali REGHEZZA-ZITT
École normale supérieure – PSL
LGP-Meudon (UMR 8591)
Paris
France

Sandrine REVET
CERI
Sciences Po – CNRS
Paris
France

Samuel RUFAT
Institut universitaire de France
Paris
France

Index

S

T, U

V

Printed and bound by CPI Group (UK) Ltd, Croydon, CR0 4YY
13/09/2026
14961046-0001